THE BIOPHYSICAL APPROACH
TO EXCITABLE SYSTEMS

Kenneth S. Cole

THE BIOPHYSICAL APPROACH TO EXCITABLE SYSTEMS

A Volume in Honor of Kenneth S. Cole
on His 80th Birthday

Edited by

William J. Adelman, Jr.

and

David E. Goldman

Laboratory of Biophysics, NINCDS
National Institutes of Health
At the Marine Biological Laboratory
Woods Hole, Massachusetts

PLENUM PRESS • NEW YORK AND LONDON

Library of Congress Cataloging in Publication Data

Main entry under title:

The biophysical approach to excitable systems.

Bibliography: p.
Includes index.
1. Excitation (Physiology) — Addresses, essays, lectures. 2. Neural conduction — Addresses, essays, lectures. 3. Biological transport, Active — Addresses, essays, lectures. 4. Biophysics — Addresses, essays, lectures. 5. Cole, Kenneth Stewart, 1900-   . I. Cole, Kenneth Stewart, 1900-   . II. Adelman, William J., 1928-   . III. Goldman, David Eliot, 1910-

QP363.B56	574.19'12	81-15759
ISBN-13:978-1-4613-3299-2	e-ISBN-13:978-1-4613-3297-8	AACR2
DOI: 10.1007/978-1-4613-3297-8		

Preface

On July 10, 1980, Kenneth S. Cole became 80 years old. In order to celebrate this landmark, a symposium in the form of a series of Monday evening lectures was held in his honor at the Marine Biological Laboratory throughout the summer of 1980. The selection of speakers was made from among those investigators who had been either his students or co-workers.

One intent of the symposium was to examine the current status of knowledge of those areas of interest in excitable membrane structure and function that owe their initiation or encouragement to Kacy Cole. The papers assembled in this volume represent a large majority of the presentations given during the 1980 Cole Symposium.

It seems clear on examination of these papers that Kacy's interests in membrane impedance, ion channel conductances, channel fluctuation phenomena, excitation, and the development of membrane biophysical methodology are all being actively pursued. It is also clear that many of his suggestions have borne fruit. Of these, his invention of the voltage clamp method has been most productive. It is hoped that these papers will provide new directions for investigations into the nature of excitable membrane phenomena.

The organizers of the symposium and the editors of this volume wish to express their thanks to the Marine Biological Laboratory for making available the facilities for the symposium. They also wish to thank Dr. Harry Grundfest for his cooperation and for having founded the Monday evening "Electrobiology" seminars at the MBL which were the forum for Kacy's 80th birthday symposium.

We thank also Ms. Dorothy Leonard for her unflagging assistance in preparing this book for publication, and Mr. Robert Golder for his artistic portrait of Dr. Cole, which is the frontispiece of this book.

Lastly, we the authors and editors of this volume wish to express our profound gratitude to Kacy for the influence he has had on our investigations. We hope that he has many more birthdays and that he can continue "membrane watching," as we believe this will continue to be both interesting and informative.

Marine Biological Laboratory   William J. Adelman, Jr.
Woods Hole, Massachusetts   David E. Goldman

Contents

PART II. MEMBRANE CHANNELS

PART III. MEMBRANE TRANSPORT

Part I

Electrical Characteristics of Membranes

Electrical Properties of Cells: Principles, Some Recent Results, and Some Unresolved Problems

H. P. SCHWAN

My own interest in the electrical properties of biological systems was largely motivated by Cole. As a young physics student, financial problems forced me to interrupt my studies until I found employment as an electronics technician at the present Max Planck Institute for Biophysics. I must confess that I was not particularly enthusiastic at first. How could it be possible to apply physics to such complex systems as exist in biology? At that time Rajewsky gave me some twenty reprints by Cole and Fricke. Here I found that the choice of appropriate models and rigorous calculations, coupled with experimental experience, patience, and ingenuity, could indeed penetrate into problem areas previously considered hopeless and derive significant results. I met Cole for the first

H. P. SCHWAN • Max Planck Institut für Biophysik, Frankfurt a. M., Germany, and Department of Bioengineering, University of Pennsylvania, Philadelphia, Pennsylvania 19104 (permanent address).

time ten years later and since then have benefited from his ideas and encouragement, as did so many of his students and collaborators.

The electrical properties of cells and tissues have been of interest to scientists ever since techniques for resistance and impedance determinations became available. Before 1910, Hoeber (1910, 1912) had already demonstrated with an elegant experiment that very high frequencies, at that time available only as dampened oscillations, yield different conductivity values for erythrocytes than low frequencies. He argued convincingly that the interior of erythrocytes is highly conducting and surrounded by a membrane. To my knowledge, he was the first to derive the existence of biological membranes from an electrical experiment. His experiments demonstrated the potential power of electrical currents to decipher the complexity of biological organization. But it was in the late Twenties and Thirties that Cole and Fricke first used advanced instrumentation and theory to analyze quantitatively a number of cellular systems. (For a summary and references, see Cole, 1972.) Since then, a very large number of investigations have been carried out on cell suspensions, tissues, and subcellular organelles, such as mitochondria, vesicles, and membranes. In addition, a large number of biopolymers have been studied, including proteins, amino acids, and nucleic acids. As a consequence of this extensive effort, utilizing ever-greater sophistication in instrumentation and analysis, detailed insight has been gained into the mechanisms which determine the dielectric properties of biological systems. These mechanisms in turn contribute to our understanding of the interaction of electrical fields with biological systems.

Practical motivations for this extensive effort were provided by the development of more efficient ultrashortwave and microwave therapeutic techniques, electrocardiography, impedance plethysmography, and concern about health hazards resulting from exposure to nonionizing electromagnetic fields. In all such cases, knowledge of tissue properties is a prerequisite to the development of relevant principles. More fundamental is the interest of the biologist and biophysicist. Membrane properties can be deduced from the electrical properties of cells and the properties of tissue water extracted. The complex structure of muscle and other tissues reveals itself in characteristic frequency dependences (Schwan, 1957), and, more recently, the structure of some cellular entities such as the gall bladder epithelium have been studied using ac impedance techniques (Schifferdecker and Froemter, 1978). And, last but not least, there is simply a desire to understand the almost beautiful and, at first glance, so simple, ac spectroscopic properties of cells for their own sake.

In alternatihg current spectroscopy, linear electrical properties are measured as a function of the frequency of the applied alternating signal. And linear electrical properties are defined as properties measured with signal levels sufficiently small so that the measured quantities are independent of signal strength. For biological materials these linear properties are conductivity σ and permittivity or dielectric constant ε, representing conductance and capacitance of a volume unit of matter.

One frequently observes characteristic frequency dependences or "dispersions" of the type (Figure 1)

$$\varepsilon, \sigma = a + \frac{b}{1+x^2} \tag{1}$$

where x is a normalized frequency, $x = f/f_0$, and the characteristic frequency

$$f_0 = \frac{1}{2\pi\tau} \tag{2}$$

where τ is a time constant. These dependences are predicted if one

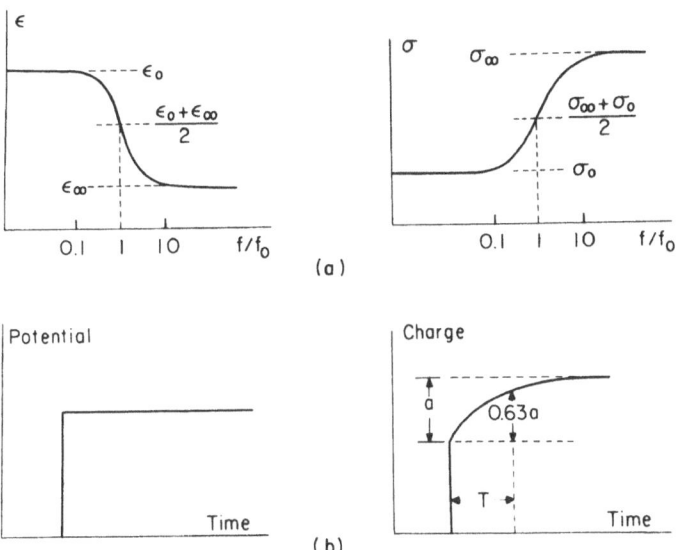

Figure 1. Frequently observed dispersion of dielectric constant ε and conductivity σ is indicated in (a). The corresponding behavior in the time domain is shown in (b), where the accumulated charge changes first instantly and then gradually towards a steady state value as a step potential is applied.

assumes an exponential response of charge or current in an experiment where a step signal is applied. The same time constant characterizes the relaxation of the sample to its original state if the step potential or current is removed.

The approach chosen is then as follows: measure dielectric constant and conductivity as a function of frequency, approximate by best fit the observed results by a single or several relaxation functions, and find out what mechanisms are responsible for these relaxation mechanisms.* Advantage is frequently taken of the fact that conductance changes and changes of dielectric constant are interrelated by the Kramers–Kronig relationships. (For a relevant discussion, see Bottcher, 1952.) For example, for one relaxation mechanism

$$\frac{\varepsilon_1 - \varepsilon_2}{\sigma_1 - \sigma_2} = \tau \tag{3}$$

with the subscripts indicating two different frequencies, f_1 and f_2.

In Figure 2 we present a schematic diagram which summarizes the dielectric constant of muscle as a function of frequency over a very wide frequency range. Two features stand out. First, the dielectric behavior is characterized by three distinct dispersions, α, β, and γ. And second, the permittivity reaches enormous values, approaching four million as the frequency decreases below 100 Hz. These amazingly high permittivity values are, to my knowledge, unsurpassed.

The origin of the rf relaxation behavior (β dispersion) was clarified about five decades ago by Cole and Fricke. In their work with a variety of cellular systems they applied concepts originally advanced by Maxwell (Cole, 1972) for direct current (dc) and by Wagner (Schwan, 1957) for alternating current (ac). They demonstrated that the dispersion effect is caused simply by the charging of the capacitive cell membranes through cytoplasmic and extracellular fluids. Since it takes time to completely charge the membranes, rapidly alternating fields are less able to accomplish this charge than are more slowly alternating ones. Cole recognized that while it is possible to extract membrane capacitance values from the

---

*The approach is based on the assumption that in the "time domain," responses are only exponential. This need not be the case. For example, diffusion controlled processes proceed nonexponentially. But, fortunately for us, the mechanisms responsible for the dielectric properties of biological systems are largely exponential.

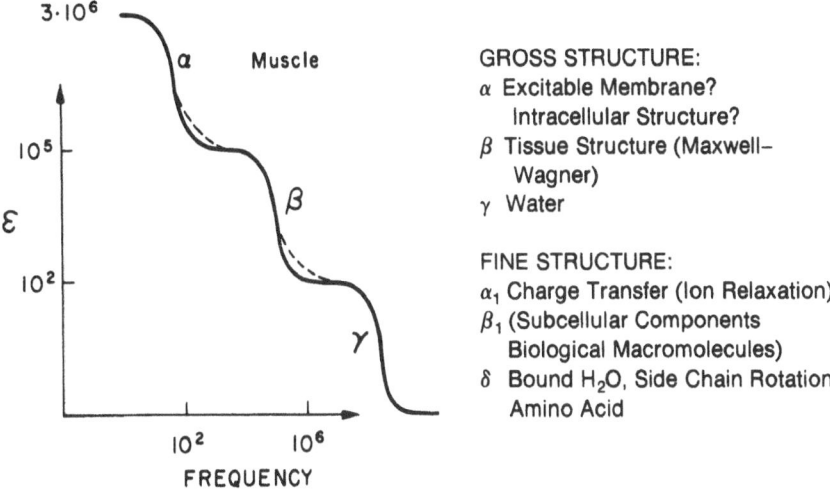

Figure 2. Dielectric constant of muscle tissue as a function of frequency. Three dispersions α, β, and γ occur at low, medium, and very high frequencies. Gross and fine-structure effects are indicated and explained in the text. The deviations of experimental data (dashed) from simple relaxational behavior (solid curve) are caused by subsidiary dispersion effects indicated by α_1, β_1, and δ as discussed in the text.

bulk properties of cell suspensions, it is almost impossible to obtain good data for the membrane conductance. But he and Curtis nevertheless succeeded with extracellular electrodes in demonstrating that strong membrane conductance changes occur upon excitation (Cole, 1972). This observation provided the stimulus for the extended effort in axonology so well known.

Work concerned with the dielectric properties of cell suspensions and tissues came almost to a stop and was not resumed again until the 1950s. A number of problems had remained unfinished, even though the mechanism responsible for the β dispersion had been clarified. For example, the experimental data for the conductivity of cytoplasmic fluids had really not been well compared with data derived from known concentrations of ionic species. And why was it that the cell membrane capacitance behaved in some instances like a perfect frequency-independent capacitor and in others like a constant phase element as exists at the interface between a metal electrode and an electrolyte? And could the concepts which emerged from the earlier work with some cell suspensions be extended to subcellular organelles, bacteria, and a variety of other biological entities such as pleuropneumonia–like organisms, virus particles, all sorts of vesicle systems, etc.? We decided in the early 1950s to extend the observations in the rf range and to substantially extend the range of observation to very low frequencies and microwave frequencies. The development of electronics during the war and thereafter permitted us to develop techniques, particularly at very low and microwave frequencies, of previously unattained accuracy (Schwan, 1963). I cannot summarize this total effort in this paper. I shall concentrate on two topics, one still in part unresolved and the other, in my opinion, satisfactorily clarified only recently.

LOW-FREQUENCY RELAXATION MECHANISMS

The dispersion of conductivity and dielectric constant in the α range is often characterized by either a single or a small distribution of time constants. But the magnitude of the dispersion of the dielectric constant ranges from nothing to several millions. Some examples are given in Table I in terms of equivalent cell membrane capacitance changes with frequency. The following mechanisms have been suggested:

1. A Sandwich Type of Membrane Structure (Schwan, 1954). The membrane model assumes two layers of proteins and a central lipid layer (Figure 3). This model and its corresponding circuit indicated in Figure 3 display an overall frequency dependence of its total capacitance and conductance characterized by a single time constant. Dispersion effects

Table I. Typical Data which Characterize the α Dispersion of Several Biological Systems[a]

	ΔC	f_c
E. coli	$>10\,\mu F/cm^2$	3 kHz
Muscle	30	0.1
Ghosts	0.6	2.5
PPLO	0.1	100

[a]Characteristic frequencies f_c are given and apparent membrane capacitance changes $\Delta C = C_0 - C_\infty$ with frequency over their dispersion range as calculated from equation (8). These apparent membrane capacitance changes are not necessarily caused by the membrane per se since equation (8) neglects the existence of several effects taking place outside the membrane as discussed in the text.

Figure 3. A membrane consisting of a lipid layer (L) covered by two protein layers (P) is assumed and modeled by the network in the center. The two protein layers are combined into one network, assuming these two layers are identical. The equivalent network on the right, composed of frequency-independent components, displays dispersion behavior with a time constant $T = \dfrac{C_L + C_P}{G_L + G_P}$.

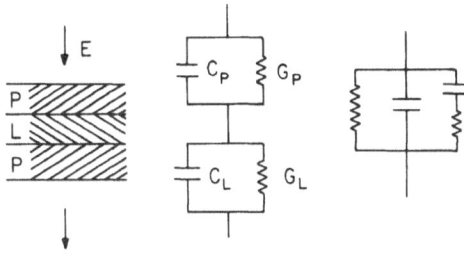

are anticipated to occur at low frequencies for typical membrane conductivities and capacitances (Schwan, 1954). But while this membrane model may have been acceptable in the Fifties, it is no longer so today. A simple double layer structure is not representative of the distribution of proteins through the membrane structure and membrane conductances cannot be individually assigned to the two layers as shown in Figure 3. They occur instead at specific sites and are in parallel to the total membrane structure.

2. A "Polarization" Element in Series with the Membrane. A boundary potential exists in the immediate neighborhood of the membrane which will be modulated by an alternating current. The modulation potential is proportional to the current density and thus defines an impedance element in series with the membrane. A double layer capacitive component of more than 10 μF/cm^2 can be estimated. This capacity, rather than the membrane capacitance, will determine the apparent capacitance of the total system at frequencies low enough to make the capacitive membrane susceptibility small compared to the membrane conductance. This happens at frequencies f so that $2\pi f < 1/\tau$, where τ is the membrane time constant $\tau = R_m C_m$ (R_m, and C_m are, respectively, membrane resistance and capacitance per 1 cm^2 area). However, little attention has been directed so far to this model.

3. Surface Conductance and Counterion Relaxation. Cells possess fixed charges and, hence, a counterion atmosphere. Electrostatic forces between fixed charge and counterion prevent movement perpendicular to the membrane surface but not tangential to it. Hence, counterions are mobile in the direction tangential to the membrane surface. This mechanism causes the apparent surface conductance to change with frequency and, hence, as demanded by the Kramers–Kronig relationships, to contain an imaginary part. We were able with high-resolution equipment to demonstrate that these changes are of the relaxational type (Schwan et al., 1962). They can be well described by the Cole–Cole type relaxation equation

$$Y_s = Y_\infty + \frac{Y_0 - Y_\infty}{1 + (j\omega\tau)^\alpha} \qquad (4)$$

where Y_s is the surface admittance

$$Y_s = G_s + j\omega C_s \qquad (5)$$

with G_s surface conductance and C_s surface capacitance and the subscripts 0 and ∞ indicating limit values for low and high frequencies compared to the characteristic frequency $f_c = 1/2\pi\tau$ and τ the average time constant of the relaxation process. Rather abundant data with polysterene latex particles of various sizes and fat particle emulsions had been collected (Schwan, 1954; Schwan and Maczuk, 1959; Schwan et al., 1962). But the counterion theory which explained the frequency-dependent contribution to the surface conductance was first provided later by Schwarz (1962). The theory could only provide closed form solutions for the dielectric constant

$$\varepsilon_0 = \varepsilon_\infty + \frac{e^2\sigma r}{kT_0}\frac{9}{4}\mathbf{p} \qquad (6)$$

and the time constant

$$\tau = \frac{r}{2\mu kT} \qquad (7)$$

with e the elementary charge, σ the charge density, r the particle radius, k the Boltzmann constant, T_0 the absolute temperature, $\mathbf{p}$ the particle volume fraction, μ the ion mobility, if it was assumed that the extent of the counterion atmosphere is small compared with the radius of the particle. This is true for biological cells. It was also assumed that the effective dielectric constant of the particle makes only a negligible contribution to the total polarization. This latter assumption is unfortunately not correct for cells since the effective dielectric constants of biological cells are very high at low frequencies, reaching values of many thousands relative to free space. Thus, the theory of Schwarz is of limited usefulness for biological cells and needs to be extended. This mechanism was suggested as possibly responsible for the high dielectric constants of muscle tissue at a time when a theoretical treatment of this effect was not yet available (Schwan, 1957). Later, Fatt confirmed the relaxation mechanism, but pointed out that the counterion hypothesis would require unusually high fixed charge densities (Falk and Fatt, 1964).

   4. Intracellular Organelles Connecting with Outer Membrane Surface. Access to a large membrane system such as the endoplasmic reticulum may be modeled as a resistor in series with the folded membrane surface. Model calculations demonstrate that the effect occurs at low frequencies. Falk and Fatt (1964) suggested that the frequency-dependent access to the tubular system is responsible for the very high dielectric constants of muscular tissue. The change from an apparent membrane capacitance of $1 \ \mu F/cm^2$ to $30 \ \mu F/cm^2$ (Schwan, 1954) as the frequency changes from 1 kHz to 10 Hz would be explained by a total membrane system with a frequency-independent value of $1 \ \mu F/cm^2$ if the subcellular membrane area is 30 times larger than that of the outer membrane. Supporting evidence of this model was provided by Freygang et al. (1967).

   5. Membrane Relaxation Effects. It is readily shown from a linearization of the Hodgkin–Huxley equations that relaxation effects of the dielectric properties of the membranes can be expected at low frequencies. Such changes in membrane properties must proportionally reflect in low-frequency dielectric properties since (Schwan, 1957)

$$\varepsilon_l = \varepsilon_\infty + \mathbf{p} r C_m \sim \mathbf{p} r C_m \qquad (8)$$

(ε_l is the dielectric constant at low frequencies, ε_∞ is the dielectric constant at very high frequencies, $\mathbf{p}$ is the volume fraction of cells, r is the cell radius, and C_m is the membrane capacitance per unit area).*

   The perplexing variety of α-dispersion phenomena observed with biological cells was already indicated in Table I. Figure 5 attempts to summarize some of these results. Erythrocytes apparently do not display

---

*Usually ε_∞ can be neglected and, hence, ε_l becomes proportional to C_m. The equation is based on spherical cellular shape, but the proportionality of ε_l and the apparent membrane capacitance C_m has general validity.

an α-relaxation behavior at all (Schwan et al., 1954), but red cell ghosts relax near 2.5 kHz (Schwan and Carstensen, 1957) and the low-frequency limit of the membrane capacitance is about 1.5 times larger than its high-frequency value of about 0.9 $\mu F/cm^2$. The strongest α-dispersion effect observed so far is that of muscle tissue, with the dielectric constant increasing from 10^5 to 3.10^6 with a relaxation frequency of only 80 Hz (Schwan, 1954). Other tissues with high water content display similar large increases (Schwan and Kay, 1957). Bacteria also display large increases, ranging from values near 10^4 to values of about 10^5 (Schwan, 1957). But their low-frequency behavior appears to be characterized by a rather broad spectrum of relaxation times. PPLO (pleuropneumonia–like organisms) have a rather small α dispersion occurring at a rather high relaxation of about 100 kHz (Schwan and Morowitz, 1962). Several vesicle systems have been investigated (Schwan and Morowitz, 1962; Schwan et al., 1970), with one displaying α-dispersion and the other not. And, finally, the squid axon membrane has been reported to display dispersion near 2.5 kHz (Takashima and Schwan, 1974) by an amount similar to that of the erythrocyte ghost (Schwan and Carstensen, 1957).

Table II indicates which phenomena have been explained in terms of some of the models outlined before. Carstensen and his co-workers (Carstensen et al., 1965, 1967, 1968; Einolf and Carstensen, 1969) have done extensive work on bacteria and protoplasts, and extended the Schwarz theory (Einolf and Carstensen, 1971) and applied it with due consideration of the large amount of fixed charges in the bacterial wall (Einolf and Carstensen, 1973). They have convincingly argued that the low-frequency effects in bacteria are due to the counterion displacement mechanism. The same is true for the vesicle system investigated by Schwan et al. (1970). Electrophoretic measurements on these vesicles were conducted and a surface charge determined which explains the

Table II. Summary of Some α-Dispersion Studies as Discussed in the Text and Suggested, or Established, Dispersion Mechanisms

Bacteria	Carstensen et al.	Counterion
Vesicles	Schwan et al.	Counterion
PPLO	Schwan et al.	Counterion (?)
Muscle	Schwan	Counterion
Muscle	Schwan	Double layer (?)
Muscle	Fatt	Organelles
Eryth. Ghosts	Schwan and Carstensen	Counterion (?)
Squid Axon	Takashima et al.	Membrane

magnitude of the dispersion using equation (6), and the relaxation frequency fits that calculated from equation (7). The high relaxation frequency simply reflects the small size of the vesicles. The same calculation has been carried out for the PPLO, investigated by Schwan and Morowitz (1962). The characteristic frequency fits fairly well, but the magnitude could not be calculated since we did not know the surface charge.

Counterions had originally been suggested as possibly responsible for the α dispersion of tissues (Schwan, 1957) at a time when the counterion theory had not yet been formulated. But this explanation cannot be sustained since equation (6) would predict a magnitude much smaller than observed if typical charge densities are about one elementary charge per 10 Å squared area. This was pointed out by Fatt (1964), who convincingly argued that a frequency-dependent access to the tubular system is responsible (Falk and Fatt, 1964; Fatt, 1964). More recently it has been speculated that in many tissues a fuzzy "greater" membrane structure may contain filaments extending from the membrane and with high fixed charge densities (Adey, 1977). This would demand a corresponding large counterion atmosphere. The counterion model developed by Einolf and Carstensen (1971) may then be applicable. The relative contribution of counterions and the tubular system (and other organelles such as the reticulum) to the α dispersion of muscle and other tissues remains unresolved.

In the case of erythrocyte ghosts the dispersion cannot be due to subcellular organelles connecting with the outer membrane since such organelles are not known to exist. The magnitude of the dispersion could be explained by a reasonable fixed charge density using equation (6). But the relaxation frequency calculated from equation (7) is much too low compared with the experimental value of 2.5 kHz.

While measuring the bulk dielectric properties of tissues and cell suspensions, fairly uniform fields are applied over cellular dimensions. This causes counterion displacement in the field direction and results in surface admittance dispersion effects as discussed above. Fields in a radial direction can be achieved with internal–external electrode pair arrangements. Such fields are less likely to move counterions in a radial direction since counterions interact with the fixed cellular charges by strong Coulombic forces. We decided, therefore, to employ radial fields and investigate the membrane of the giant squid axon in order to exclude counterion effects from contributing to membrane relaxation effects.

H. P. Schwan

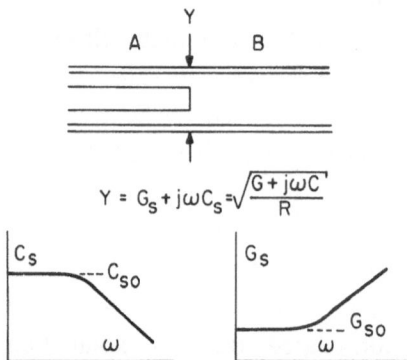

$$Y = G_s + j\omega C_s = \sqrt{\frac{G + j\omega C}{R}}$$

Figure 4. The stray field of a metal electrode inside a squid axon is modeled. In the upper portion of the figure the axon with the metal electrode inserted is illustrated on the left; the axon free of the electrode is illustrated on the right. It joins with the left section at Y. The equation presents the input admittance Y of the right section which terminates the left section. The frequency dependence of its two components, conductance G_s and capacitance C_s, is indicated in the lower part. G and C are membrane conductance and capacitance per unit length; R is axon resistance per unit length.

The relaxation observed with the squid axon membrane (Takashima and Schwan, 1974; Takashima, 1976) was probably, at least in part, due to stray field effects. Later attempts to compensate for the stray field by variation in electrode length resulted in a reduction of the dispersion amplitude. It is possible to estimate the stray field effect. Figure 4 indicates the approach. The stray field admittance is equal to the input admittance of a transmission line section without the central electrode, loading the axon with the electrode at the electrode end. This input admittance is

$$Y_0 = G_s + j\omega C_s = \left(\frac{G + j\omega C}{R} \right)^{1/2} \tag{9}$$

with G_s and C_s conductance and capacitance of the stray field admittance; and C, G, R capacitance, conductance, and resistance, respectively, per unit length of the stray field section; and $\omega = 2\pi f$ angular frequency. The inductance per unit length is neglected as usual in axon problems at low frequencies. From this we have

$$G_s^2 = \frac{G}{2R} \left\{ 1 + \left[1 + \left(\frac{\omega C}{G} \right)^2 \right]^{1/2} \right\} \tag{10}$$

$$(\omega C_s)^2 = \frac{G}{2R} \left\{ -1 + \left[1 + \left(\frac{\omega C}{G} \right)^2 \right]^{1/2} \right\} \tag{11}$$

Limit values at low and high frequencies are

$$G_{s_0}^2 = \frac{G}{R}, \qquad G_{s_\infty}^2 = \frac{\omega C}{2R} \to \infty \qquad (12)$$

$$C_{s_0} = \frac{C}{2(RG)^{1/2}}, \qquad C_{s_\infty} = \left(\frac{C}{2\omega R}\right)^{1/2} \to 0 \qquad (13)$$

C_{s_0} may be rewritten to demonstrate the effect of radius r

$$C_{s_0} = \frac{C}{2}\left(\frac{r}{2\rho G_m}\right)^{1/2} \qquad (14)$$

with ρ the resistivity of the internal medium and G_m the axon's membrane conductance per unit area. For typical values $G_m = 10^{-3}$ mho/cm^2, $\rho = 20$ Ωcm, $r = 400$ μm, the stray field capacitance at low frequencies is $C_{s_0} = 0.5$ C, i.e., the stray field contributes 50% to the membrane capacitance per centimeter. The frequency needed to reduce C_{s_0} by a factor of 2 is given by

$$C_s = \frac{1}{2}C_{s_0} = \frac{C}{4(RC)^{1/2}} \qquad (15)$$

which yields from equation (11)

$$\left(\frac{\omega C}{G}\right)^2 = 48 \to \frac{\omega C}{G} \sim 7 \qquad (16)$$

The frequency needed to double the value of G_{s_0} is readily derived to be the same. For $C_m = 1$ μF/cm^2 and $G_m = 10^{-3}$ mho/cm^2, $\omega = 7000$, i.e., the frequency f is near 1.1 kHz. The following results are readily derived:

$$\frac{G_{s_0}}{G} = \frac{1}{(RG)^{1/2}}, \qquad \frac{C_{s_0}}{C} = \frac{1}{2(RG)^{1/2}} \qquad (17)$$

i.e., G_{s_0} compares with G since $C_{s_0} \sim 0.5$ C for a radius $r = 400$ μm. The above analysis tacitly assumes that the internal electrode has a diameter equal to that of the axon. If the electrode diameter is smaller than that of the axon, the stray field may deviate somewhat from the one calculated and a more detailed analysis is required. But the corrections needed in the above equations should not be serious since the internal electrode will enforce a constant potential along its length inside the axon membrane and the radial potential change in the highly conducting interior is small. We conclude that the stray field capacitance and conductance constitute appreciable fractions of the axon capacitance and conductance per centimeter of length. This effect diminishes rapidly above about 1 kHz and thereby may well simulate a dispersion effect of the membrane capacitance.

Attempts to correct for the stray field were made by varying the length of the internal electrode and noting the differences between the results obtained with different lengths (Takashima, 1976). This technique should eliminate stray field contributions provided they do not change due to a changing diameter of the axon. A dispersion effect was observed, of smaller magnitude than originally stated. Fishman also obtained results which support the existence of the relaxation effect near 2.5 kHz (personal communication). More work is indicated to entirely resolve the problem of possible stray field contributions to relaxation effects in membranes.

We have not discussed the frequency dependencies which may result from an ion gating mechanism. We already stated that the Hodgkin–Huxley equations suggest such effects at the linear level. In recent years Fishman and his colleagues have studied the squid axon membrane at frequencies extending substantially below 1 kHz (Fishman et al., 1977, and personal communication; Poussart, 1977). Strong frequency dependencies are observed, which appear to correlate well with the predictions of the linearized Hodgkin–Huxley equations. De Felice and his colleagues have studied the squid axon membrane admittance as a function of frequency and clamped dc potential with small ac signals. They report sharp resonantlike admittance peaks for certain combinations of clamped dc potential and frequency (personal communication), and second-order admittances generating harmonics and suggesting nonlinear admittance components at ac potentials as small as 1 mV across the membrane (De Felice et al., 1980, and personal communication). As stated in the introduction, this summary is entirely restricted to dielectric relaxation effects at the linear level.

In summary: So far no membrane system has been identified with certainty which displays dielectric relaxation behavior directly at frequencies above 1 kHz, except for some observations reported above. Counterion relaxation is responsible in many cellular suspensions for the observed low-frequency relaxation effects. But the amazingly high static dielectric constants of tissues may be caused primarily by subcellular organelles. Counterion contributions provide an alternate explanation only if very high charge densities in a "greater" membrane are assumed. More work is needed to extend the Schwarz theory of counterion relaxation so that it can be applied with more confidence to biological cells. To what extent ion gating mechanisms contribute to dielectric relaxation mechanisms is presently under investigation.

THE STATE OF TISSUE WATER

As mentioned before, Hoeber had already obtained a value for the conductance of the red cell interior. The work of Fricke and that of Cole provided additional data with no indication that tissue water might be differently structured from normal water. Earlier data frequently had to rely on extrapolations to higher frequencies where membrane properties no longer affect dielectric properties. But such extrapolations are difficult since tissue properties in the rf range include secondary dispersion phenomena caused by mitochondria, cell nuclei, and biopolymers. A direct and more precise determination of internal conductance values requires frequencies near 100 MHz, high enough to short-circuit and thereby eliminate membrane contributions and yet low enough to exclude conductance contributions from the rotation of water molecules. We measured the dielectric properties of various tissues and cell suspensions over the frequency range from 100 to 1000 MHz (Schwan and Li, 1953; Schwan, 1957). The results may be described as follows:

(1) The dielectric constant of tissues of high water content and blood are identical to that of normal water if allowance is made for the space occupied by tissue proteins and hemoglobin.

(2) The conductance values are approximately comparable to those calculated from the concentrations of ionic species. But there appears to be a systematic deviation with experimental values smaller than calculated ones. A detailed study was conducted with erythrocytes and ghosts containing varying amounts of Hb (Pauly and Schwan, 1966). In the ghosts agreement between theory and experiment was achieved at Hb concentrations less than half normal. At the higher concentration level

internal friction with proteins reduces the mobility of ions noticeably. This effect, if adequately considered, entirely removes the discrepancy between theory and experiment. No comparable analysis has yet been conducted for tissues and other cell suspensions.

(3) Dielectric work on protein suspensions over the range from 100 to 1000 MHz (Schwan, 1957, 1965; Grant, 1965; Pennock and Schwan, 1969) indicated a small fraction of protein bound water, about 0.3 g bound water per gram of protein. This protein bound water or polar side chain rotation causes the effective dielectric constant of the hydrated protein to change with frequency, with relaxation frequencies extending from a few hundred to a few thousand megahertz. However, the dielectric and conductive contributions of this relaxation behavior on the total dielectric properties of tissues and cell suspensions are not very pronounced.

Dielectric data for tissues and blood were also available for the frequency range from 1000 to 8500 MHz. These data strongly indicated that the relaxation frequency for tissue water and normal water are identical even though the range of observation did not extend to the relaxation frequency of normal water.

Thus it appears that dielectric constants near 100 MHz, ion mobilities obtained from the conductance data, and tissue water relaxation times compare well with properties characteristic of normal water if we exclude the small volume taken by protein-bound water. However, this point of view has been challenged. Ling and his colleagues (1973) stated that tissue water is differently structured from normal water, a view shared by a small but active group of biophysicists. Drost-Hansen (1977) also believes that the identification of tissue water with normal water is naive and has developed his own concepts of structured water. Carpenter and co-workers (Hovey *et al.*, 1972) attempted to determine the internal conductance of aplysia cells and derived values which simply could not be understood unless one assumed vastly different properties of cellular and normal water. However, in a subsequent paper (Foster *et al.*, 1976), this earlier work was corrected and the original internal conductance values were replaced by values supportive of our views stated above. Then there appeared a series of papers by Masszi and his colleagues (Masszi, 1972; Masszi *et al.*, 1976) stating that the relaxation frequency of tissue water is appreciably shifted from that of normal water. Their work was carried out at a frequency of 2.45 GHz and based on an evaluation of conductance data. At this frequency the Debye contribution to the low-frequency conductivity is already appreciable and sensitive to the value of the relaxation frequency. But in a more recent detailed analysis of available tissue data we could not find a confirmation of this conclusion (Schwan and Foster, 1977).

Techniques have recently become available in our laboratory which extend the range of observation to 18 GHz, close to the relaxation frequency of normal water at room temperature and twice that of 9°C water. We decided to confirm the previously unpublished data by Herrick which had been made available to me, to extend tissue data to 18 GHz, and to review the question of tissue water with the precision microwave techniques now available using a microwave network analyzer. The dielectric properties of barnacle muscle were investigated first, concentrating on the relaxation frequency of tissue water (Foster *et al.*, 1980). Two approaches were chosen and are illustrated in Figures 5 and 6. Figure 5 presents the Cole–Cole circle. The static dielectric constant of 62 is as one might predict using Maxwell's mixture equation to account for the protein volume. The peak of the circle illustrates the relaxation frequency. It is identical to that of free water. The second approach evaluates conductance data alone and is an adaptation of a technique used by us before (Schwan *et al.*, 1976). The conductance is plotted versus a normalized conductance axis as indicated. The result is a straight

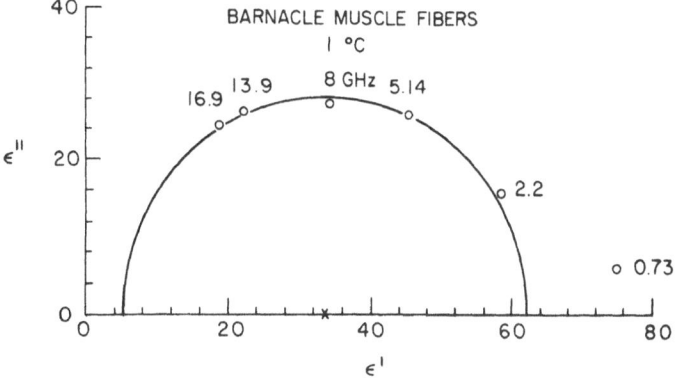

Figure 5. Cole–Cole circle for barnacle muscle. The peak frequency of 8 GHz is the characteristic frequency of the dispersion f_c and is equal to that of normal water at 1°C. [From Foster *et al.* (1980), with permission of the authors.]

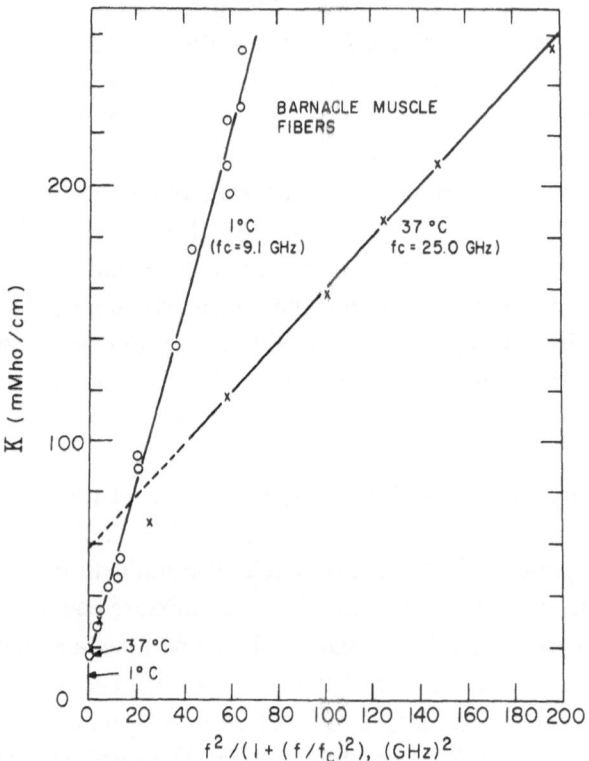

Figure 6. The microwave conductivity of barnacle muscle as a function of the normalized frequency square $f^2/[1+(f/f_c)^2]$. The theory demands a line of constant slope. From this slope the relaxation time for tissue water is identical to that of normal water. Deviations at lower frequencies and a temperature of 37°C are caused by the additional uhf relaxation mechanism typical for protein solutions and tissues. [From Foster *et al.* (1980), with permission of the authors.]

line, as may be seen from the relaxation equation

$$\sigma = \sigma_0 + (\sigma_\infty - \sigma_0)\frac{(\omega\tau)^2}{1+(\omega\tau)^2} \tag{18}$$

for the conductance and its slope is given by the relaxation frequency and thus provides f_c. The technique must start with an assumption of f_c to provide the axis values. But we note that the slope of the curve is independent of this assumed value. At high frequencies values were obtained for the slope providing a relaxation frequency identical with that of normal water. But at lower frequencies the slope changes for the 37°C curve. Our analysis demonstrates that this change is caused by the subsidiary protein uhf dispersion discussed above. For a more detailed discussion see Foster *et al.* (1980). We conclude therefore that Masszi's results, obtained at much lower frequencies, were in error since he did not appropriately correct for the uhf dispersion effect unrelated to tissue water and probably caused by proteins (Schwan, 1965; Grant, 1965; Pennock and Schwan, 1969).

More recently, Jenin and Schwan (1980) have pointed out how to determine the time constant of cell water from lower-frequency data. Again, a time constant was noted which compared well with that of normal water. At the same time the dielectric properties of the hemoglobin solution inside the red cell were noted to be identical with those of a solution of hemoglobin in normal water of equal concentration.

We therefore conclude that our previous opinion of cellular water being identical with normal water, at least from a dielectric point of view, is supported by recent evidence.

In summary: It appears that our understanding of the electrical properties of cells at microwave frequencies is fairly complete. However, more work needs to be done to completely analyze the factors responsible for low-frequency properties. This is particularly important if we want to understand the mechanism of interaction of externally imposed electrical fields with tissues. One of the motivations for this article was the observation that many contributions in this field are frequently based on one model and, hence, subject to criticism. Clearly a number of relaxation mechanisms have been recognized which can contribute to the electrical properties of cells and membranes and must be properly considered in any analysis of the electrical characteristics of cells and tissues.

ACKNOWLEDGMENT

Support by the Alexander von Humboldt-Stiftung, which provided a Senior U.S. Scientist Award, and by the Office of Naval Research (N00014-78-C-0392) is gratefully acknowledged.

REFERENCES

Adey, W. R., and Bawin, S. M. (1977). Brain interactions with weak electric and magnetic fields, *Neurosci. Res. Bull.* **15**, 7.

Böttcher, C. J. F. (1952). *Theory of Electric Polarization* (Elsevier, Houston).

Carstensen, E. L., Cox, H. A., Jr., Mercer, W. B., and Natale, L. A. (1965). Passive electrical properties of micro-organisms. I. Conductivity of *Escherichia coli* and *Micrococcus lysodeikticus*, *Biophys. J.* **5**, 289.

Carstensen, E. L. (1967). Passive electrical properties of micro-organisms. II. Resistance of the bacterial membrane, *Biophys. J.* **7**, 493.

Carstensen, E. L., and Marquis, R. E. (1968). Passive electrical properties of micro-organisms. III. Conductivity of isolated bacterial cell walls, *Biophys. J.* **8**, 536.

Cole, K. S. (1972). *Membranes, Ions and Impulses* (University of California Press, Berkeley).

DeFelice, L. J., Adelman, W. J., Jr., Clapham, D. E., and Mauro, A. (1980). Second order admittance in squid axon, Abstract, presented at the 1980 Joint ASBC/Biophys. Soc. Meeting, New Orleans.

Drost-Hansen, W. (1977). Water at biological interfaces: Structural and functional aspects, *Phys. Chem. Liq.* **7**, 243–345.

Einolf, C. W., Jr., and Carstensen, E. L. (1969). Passive electrical properties of micro-organisms. IV. Studies of the protoplasts of *Micrococcus lysodeikticus*, *Biophys. J.* **9**, 634.

Einolf, C. W., Jr., and Carstensen, E. L. (1973). Passive electrical properties of micro-organisms. V. Low frequency dielectric dispersion of bacteria, *Biophys. J.* **13**, 8.

Falk, G., and Fatt, P. (1964). Linear electrical properties of striated muscle fibers observed with intracellular electrodes, *Proc. R. Soc. London Ser. B* **160**, 69–123.

Fatt, P. (1964). An analysis of the transverse electrical impedance of striated muscle, *Proc. R. Soc. London Ser. B* **159**, 606–651.

Fishman, H. M., Poussart, D., Moore, L. E., and Siebenga, E. (1977). K-conduction description for the low-frequency impedance and admittance of squid axon, *J. Membr. Biol.* **32**, 255–290.

Foster, K. R., Bidinger, J. M., and Carpenter, D. O. (1976). The electrical resistivity of cytoplasm, *Biophys. J.* **16**, 991.

Foster, K. R., Schepps, J. L., and Schwan, H. P. (1980). Microwave dielectric relaxation in muscle: A second Look, *Biophys. J.* **29**, 271–282.

Freygang, W. H., Jr., Rapoport, S. I., and Peachey, L. D. (1967). Some relations between changes in the linear electrical properties of striated muscle fibers and changes in ultrastructure, *J. Gen. Physiol.* **50**, 2437–2458.

Grant, E. H. (1965). "The structure of water, neighboring proteins, peptides and amino acids as deduced from dielectric measurements, *Ann. N.Y. Acad. Sci.* **125**, 418–427.

Hoeber, R. (1910). Eine Methode die elektrische Leitfähigkeit im Innern von Zellen zu messen, *Arch. Ges. Physiol.* **133**, 237–259.

Hoeber, R. (1912). Ein zweites Verfahren die Leitfähigkeit im Innern von Zellen zu messen, *Arch. ges. Physiol.* **148**, 189–221.

Hoven, M. M., Bak, A. F., and Carpenter, D. O. (1972). Low internal conductivity of *Aplysia* neuron somata, *Science* **176**, 1329–1330.

Jenin, P. C., and Schwan, H. P., (1980). Some observations on the dielectric properties of hemoglobin's suspending medium inside human erythrocytes, *Biophys. J.* **30**, 285–293.

Ling, G. N., Miller, C., and Ochsenfeld, M. M. (1973). The physical state of solutes and water in living cells according to the association induction hypothesis, *Ann. N.Y. Acad. Sci.* **204**, 6.

Masszi, G. (1972). Dielectric relaxation and water structure in gelatin solutions, *Acta Biochim. Biophys. Acad. Sci. Hung.* **7**, 349–357.

Masszi, G., Szuarto, A., and Grof, P. (1976). Investigations on the ion- and water-binding of muscle by microwave measurements, *Acta Biochim. Biophys. Acad. Sci. Hung.* **11**, 129–131.

Pauly, H., and Schwan, H. P. (1966). Dielectric properties and ion mobility in erythrocytes, *Biophys. J.* **6**, 621–639.

Pennock, B., and Schwan, H. P. (1969). Further observations on the electrical properties of hemoglobin bound water, *J. Phys. Chem.* **73**, 2600.

Poussart, D., Moore, L. E., and Fishman, H. M. (1977). Ion movement and kinetics in squid axon. I. Complex admittance, *Ann. N.Y. Acad. Sci.* **303**, 355–379.

Redwood, W. R., Takashima, S., Schwan, H. P., and Thomson, T. E. (1972). Dielectric studies on homogeneous phosphatidylcholine vesicles, *Biochim. Biophys. Acat* **255**, 557–566.

Schifferdecker, E., and Froemter, E. (1978). The AC impedance of Necturus gall-bladder epithelium, *Pflügers Arch., Eur. J. Physiol.* **377**, 125–133.

Schwan, H. P., and Li, K. (1953). Capacity and conductivity of body tissues at ultra-high frequencies, *Proc. I.R.E.* **41**, 1735.

Schwan, H. P. (1954). Electrical properties of muscle tissue at low frequencies, *Z. Naturforsch.* **9b**, 245.

Schwan, H. P., Bothwell, T. P., and Wiercinski, F. J. (1954). Electrical properties of beef erythrocyte suspensions at low frequencies, *Fed. Proc. Am. Soc. Exp. Biol.* **13**, 15.

Schwan, H. P. (1957). Electrical properties of tissue and cell suspensions, in *Advances in Biological and Medical Physics*, Vol. V, J. H. Lawrence and C. A. Tobias, Eds. (Academic, New York), p. 147.

Schwan, H. P., and Carstensen, E. L. (1957). Dielectric properties of membrane of lysed erythrocytes, *Science* **125**, 985.

Schwan, H. P., and Kay, C. F. (1957). Capacitive properties of living tissues, *Circ. Res.* **5**, 439.

Schwan, H. P., and Maczuk, J. (1959). Electrical relaxation phenomena of biological cells and colloidal particles at low frequencies, in *Proceedings of the First National Biophysics Conference* (Yale University Press, New Haven, Connecticut), p. 348.

Schwan, H. P., and Morowitz, H. J. (1962). Electrical properties of the membranes of the pleuropneumonia-like organism A5969, *Biophys. J.* **2**, 395.

Schwan, H. P., Schwarz, G., Maczuk, J., and Pauly, H. (1962). On the low frequency dielectric dispersion of colloidal particles in electrolyte solution, *J. Phys. Chem.* **66**, 2626.

Schwan, H. P. (1963). Determination of biological impedances, in *Physical Techniques in Biological Research*, Vol. 6, W. L. Nastuk, Ed. (Academic, New York), p. 323.

Schwan, H. P. (1965). *Electrical properties of bound water, Ann. N.Y. Acad. Sci.* **125**, 344–354.

Schwan, H. P., Takashima, S., Miyamoto, V. K., and Stoeckenius, W. (1970). Electrical properties of phosphilipid vesicles, *Biophys. J.* **10**, 1102–1119.

Schwan, H. P., Sheppard, R. J., and Grant, E. H. (1976). Complex permittivity of water at 25°C, *J. Chem. Phys.* **64**, 2257–2258.

Schwan, H. P., and Foster, K. R. (1977). Microwave dielectric properties of tissue: Some comments on the rotational mobility of tissue water, *Biophys. J.* **17**, 193–197.

Schwarz, G. (1962). A theory of low frequency dielectric dispersion of colloidal particles in electrolyte solution, *J. Phys. Chem.* **66**, 2636–2642.

Takashima, S., and Schwan, H. P. (1974). Passive electrical properties of squid axon membrane, *J. Membr. Biol.* **17**, 51–68.

Takashima, S., (1976). Membrane capacity of squid axon during hyper- and depolarization, *J. Membr. Biol.* **27**, 21–39.

2

Nonlinear Sinusoidal Currents in the Hodgkin–Huxley Model

RICHARD FITZHUGH

INTRODUCTION

Forty years ago Kacy Cole pioneered the measurement of the electric impedance of nerve membranes (Cole, 1968). In addition to a capacitative reactance from the nerve capacity, he found an unexpected inductive component. The latter could not be the result of a changing magnetic field, as is the conventional inductance in physical systems, but was interpreted as arising from changes of the potassium permeability of the membrane.

Impedance measurements are limited by the requirement of linearity. If the amplitude of the sinusoidal input current is too large, the output potential is no longer a pure sine wave, but is distorted by the

RICHARD FITZHUGH • Laboratory of Biophysics, NINCDS, National Institutes of Health, Bethesda, Maryland 20205.

nonlinearity of the conductance changes, which makes it impossible to measure the impedance accurately.

Recently, it occurred to DeFelice, Adelman, and others to consider this nonlinear distortion not merely as an obstacle in measuring linear impedances, but as a possible source of information about the nonlinear processes of excitation. I was asked to try to derive theoretically the results of experiments (DeFelice *et al.*, 1980) which they were doing to measure this effect, using the Hodgkin–Huxley (HH) model.

The method for analyzing a nonlinear system described here differs from that of Wiener kernels (Wiener, 1958), which characterizes the response of a system to a white noise input, in terms of a series of integrals of increasing multiplicity. A simplification of Wiener's method, using an input composed of the sum of a finite number of sinusoids, is described by Victor and Shapley (1980). The present method studies the response of a system to only one sinusoid at a time. The output consists of sinusoids at several frequencies, multiples of the input frequency. The number of output frequencies having appreciable amplitude increases with the amplitude of the input. For small input amplitudes, only a few output amplitudes need be considered. This procedure is then repeated at other input frequencies, instead of studying the response to several input frequencies at once, as done by Victor and Shapley. It thus forms a natural generalization of the classical linear analysis by measuring impedances.

The nerve membrane behaves like a circuit with parallel branches for the capacity and the different ionic conducting pathways. Theoretically, therefore, rather than consider its impedance, it is simpler to deal with its admittance, which is the reciprocal of the impedance, since the total membrane admittance is simply the sum of the admittances of the separate branches.

The conventional admittance can be considered as the first term in a sequence of higher-order, nonlinear, generalized admittances of a new kind. The first-order, or linear, admittance can be separated into components of the familiar *RLC* type, which help in analyzing the properties of the membrane (Chandler, FitzHugh, and Cole, 1962). It is not yet clear how to interpret the more complicated expressions which describe the higher-order admittances in an informative manner. One can be sure only that they depend on the nonlinearities of the system, and a comparison of their magnitudes will show the potential range for which the nonlinearities are largest. In particular, they are not dominated at higher frequencies by the capacitative reactance, as is the linear component.

HODGKIN–HUXLEY EQUATIONS

In the HH equations, there are three variables which determine the ionic conductances: m (the sodium activation), h (the sodium inactivation), and n (the potassium activation).

Each of these three variables obeys a differential equation of the form

$$\dot{x} = dx/dt = \phi \left[\alpha_x(V) - \gamma_x(V)x \right] \tag{1}$$

where x stands for m, h, or n. $\gamma_x(V) = \alpha_x(V) + \beta_x(V)$, where $\alpha_x(V)$ and $\beta_x(V)$ are defined as usual for $x = m, h, n$ in the HH model. ϕ is a constant equal to $3^{(T-6.3)/10}$, where T is the temperature in °C. Assume that a voltage clamp is applied, with a membrane potential of the form

$$V = V_0 + V_1 \cos \omega t, \qquad \omega = 2\pi f \tag{2}$$

where V_0 is a constant potential on which is superimposed a sinusoidal signal of amplitude V_1. After any transients resulting from the change of V_0 at the beginning of the clamp have subsided, the system reaches a steady state, and each variable m, h, n can be expressed as a power series in the amplitude V_1:

$$x(t) = x_0 + V_1 x_1 + V_1^2 x_2 + V_1^3 x_3 + \cdots \tag{3}$$

where x_0 is the steady state value of x at $V = V_0$, and x_1, x_2, x_3 are functions of time which form the coefficients of a power series for $x(t)$.

To simplify the notation, let

$$c_k = \cos k\omega t$$

$$s_k = \sin k\omega t$$

where $k = 0, 1, 2, \ldots$. Then, inserting equation (3) into (1), and solving for these coefficients, we obtain

$$x_1 = x_1^{c1} c_1 + x_1^{s1} s_1$$

$$x_2 = x_2^{c0} c_0 + x_2^{c2} c_2 + x_2^{s2} s_2 \qquad (4)$$

$$x_3 = x_3^{c1} c_1 + x_3^{s1} s_1 + x_3^{c3} c_3 + x_3^{s3}$$

These equations show that the different order components of $x(t)$ can be expressed in terms of sines and cosines, multiplied by coefficients which are functions of V_0 and f (steady potential and frequency). These coefficients are identified by a subscript which denotes the order of the nonlinearity, and a superscript ck or sk to indicate that it is to be multiplied by the cosine c_k or the sine s_k, respectively. The first-order component shows only a sine and cosine of the fundamental frequency, f, as expected. The second-order component has one term at zero frequency which is therefore constant (since $c_0 = 1$), and two more at $2f$. The third-order component has terms at both f and $3f$.

Higher-order components were not attempted, but they would be expected to follow the same pattern: odd-order components have terms at odd frequencies up to the order of that component, and even components similarly.

The total membrane current is

$$I = C\dot{V} + J(V, m, h, n)$$

$$J(V, m, h, n) = \bar{g}_{Na} m^3 h (V - V_{Na}) + \bar{g}_K n^4 (V - V_K) + g_L (V - V_L) \quad (5)$$

C is the membrane capacitance, J the total ionic current; subscript L refers to leakage. I is expanded in a power series in V_1:

$$I = I_0 + V_1 I_1 + V_1^2 I_2 + V_1^3 I_3 + \cdots \qquad (6)$$

The components of I are

$$I_0 = J_0$$
$$I_1 = I_1^{c1} c_1 + I_1^{s1} s_1$$
$$I_2 = I_2^{c0} c_0 + I_2^{c2} c_2 + I_2^{s2} s_2 \qquad (7)$$
$$I_3 = I_3^{c1} c_1 + I_3^{s1} s_1 + I_2^{c3} c_3 + I_3^{s3} s_3$$

These components follow the same pattern of frequencies as those of $x(t)$.

Detailed expressions for all these coefficients of the sines and cosines have been derived, and will be published elsewhere, but are not given here. From them can be computed curves of the nonlinear currents as a function of the steady potential level V_0, the input amplitude V_1, and the frequency f, for comparison with experiment.

The I symbols in equations (6) and (7) do not all correspond to quantities with the same physical units. I and I_0 are current densities. If they have the units of $\mu A\ cm^{-2}$, and V and V_1 have the units of mV, then the I_1's are in $\mu A\ mV^{-1}\ cm^{-2}$, the I_2's are in $\mu A\ mV^{-2}\ cm^{-2}$, and the I_3's are in $\mu A\ mV^{-3}\ cm^{-2}$.

The I_1's thus have the units of admittance per unit area. In fact, I_1^{cl} and $-I_1^{sl}$ are equal to the real and imaginary parts of the ordinary first order (linear) admittance. This is shown as follows. Let $Y=g+jh$ denote the complex admittance, where g is the conductance and h the susceptance. The complex potential is

$$V_1 e^{j\omega t} = V_1(c_1 + js_1)$$

The complex current is Y times this, or

$$V_1(g+jh)(c_1+js_1)=V_1\left[(gc_1-hs_1)+j(gs_1+hc_1)\right] \qquad (8)$$

The real part of this expression equals the $V_1 I_1$ term in equation (6). Thus, using the second of equations (7),

$$I_1 = gc_1 - hs_1 = I_1^{cl}c_1 + I_1^{sl}s_1, \qquad g=I_1^{cl}, \qquad h=-I_1^{sl}$$

The two latter quantities are plotted as coordinates of the complex plane in Figure 1 below. Although the coefficients for orders 2 and 3 do not have the same dimensions, it still seems useful to consider them as higher-order "admittances" of a new type and plot them on the complex plane, as in Figures 2–5.

Figure 1 shows a set of linear admittance loci in the complex plane for the HH model. Each curve corresponds to a different membrane potential V_0 and is plotted for positive frequencies only. (The curve for negative frequencies is obtained by reflecting the curve about the horizontal axis.) The shape of each curve is determined by the fact that the total linear membrane admittance is the sum of several terms [equation (11) and Figure 5 in Chandler et al., 1962]:

$$Y=g_\infty +j\omega c+g_m/(1+j\omega\tau_m)+g_h/(1+j\omega\tau_h)+g_n/(1+j\omega\tau_n)$$

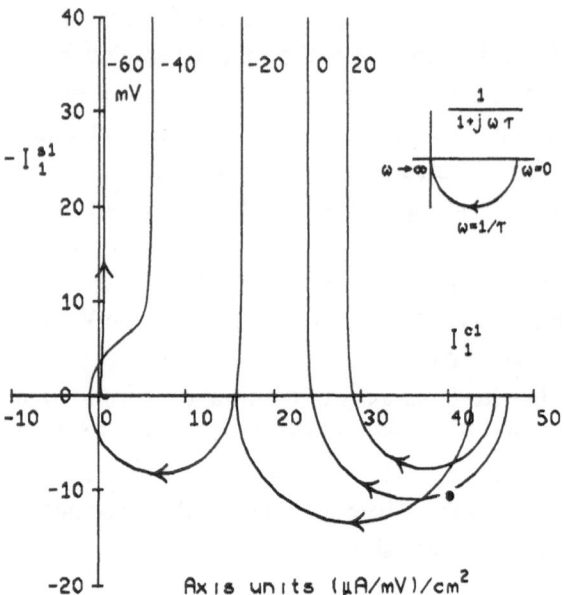

Figure 1. Linear admittance curves plotted in the complex plane for the Hodgkin–Huxley equations. For labeling of coordinates, see text. Each curve is labeled with the value of the steady potential V_0 in millivolts. The frequency starts at zero and increases along each curve in the direction of the arrowhead. Inset (at top right) shows semicircle for term of form $1/(1+j\omega\tau)$. Temperature for all figures is 6.3°C.

The g's and τ's are real functions of V_0. All of them are positive except g_m, which is negative. The first two terms plotted alone on the complex Y plane would give a vertical line to the right of the origin. The actual curve approaches this line as $\omega \to \infty$. For low frequencies, the shape of the curve is dominated by the last three terms. A term of form $1/(1+j\omega\tau)$ plotted alone in the complex plane forms a semicircle (Figure 1, inset). One end ($\omega=0$) lies on the horizontal axis at $+1$, the other end ($\omega \to \infty$) is at the origin, and the halfway point (arrowhead) is at $\omega=1/\tau$. Increasing ω corresponds to moving along the semicircle in a clockwise direction. Multiplying this curve by a positive constant magnifies it; multiplying it by a negative constant also rotates it 180° about the origin.

Since g_h and g_n are positive, their semicircles lie below the horizontal axis and to the right of the origin. Since τ_h and τ_n are fairly close together, these two semicircles cannot be distinguished in the total Y curve; together they form the bottom loop of that curve. Since g_m is negative, its semicircle lies above the axis and to the left of the origin, but only for

-40 mV is it large enough for its effect to be seen, making the curve bulge to the left and encircle the origin.

Figure 2 shows how the second-order coefficient I_2^{c0} varies as a function of frequency, for different values of V_0. I_2^{c0} represents the zero-frequency component of the second-order generalized admittance. This component produces a change of the average or dc level of the output current, which increases as the square of the input amplitude V_1.

Curves such as those in Figures 2–5 indicate the potentials V_0 for which the nonlinearities are most pronounced. In general, they are strongest for -40 mV, less so for -20 mV, and still smaller for the other potentials.

Figure 3 shows the second-order coefficients corresponding to twice the fundamental frequency; $-I_2^{s2}$ is plotted against I_2^{c2}. By analogy with the linear admittance, these two quantities can be considered as the components of a second-order generalized complex admittance. These curves, one for each potential, also curve in a clockwise fashion with increasing frequency. Unlike those in Figure 1, they do not go off to infinity, which is reasonable in the absence of a nonlinear membrane capacity.

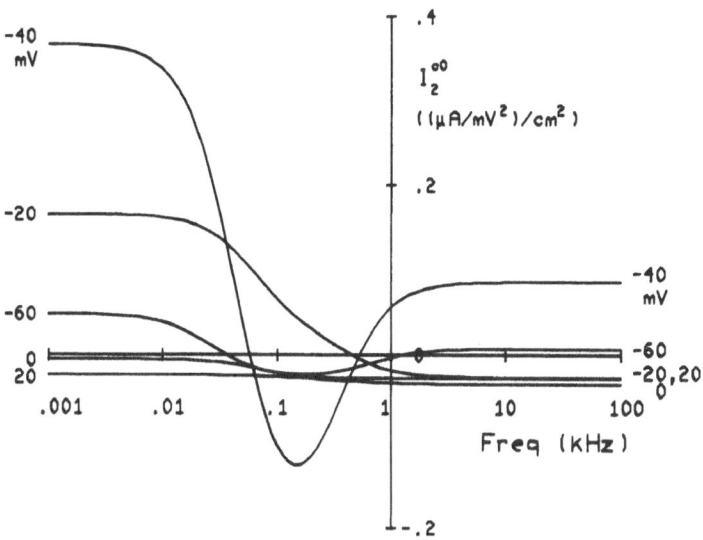

Figure 2. Second-order nonlinear generalized conductance corresponding to zero frequency, plotted against frequency for different values of V_0 (labeled).

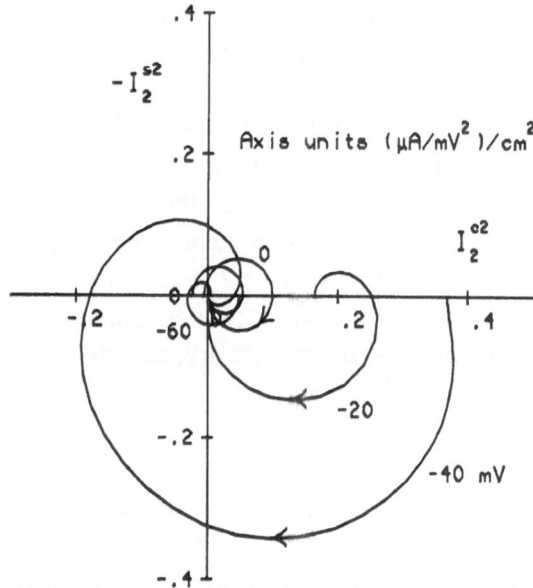

Figure 3. Second-order nonlinear generalized admittances corresponding to twice the fundamental frequency.

There are two third-order admittances, one given by the coefficient pair I_3^{c1} and $-I_3^{s1}$ at frequency f, and the other by the pair I_3^{c3} and $-I_3^{s3}$ at frequency $3f$. These are plotted on Figures 4 and 5. Notice that the scales on the horizontal and vertical axes are different in these figures and were chosen so as to make the curves more visible. When plotted with the same units on both axes, these curves appear to lie nearly along the horizontal axis.

Whether the detailed shapes of these curves for second- and third-order admittances can be interpreted in any informative way remains to be seen. For the moment they may be considered as interesting new quantities to be investigated further. One difficulty in their interpretation lies in the complexity of the mathematical expressions involved. The second-order admittances are more complicated than the first, and the third still more so.

Detailed measurements of admittance loci like those in Figures 2–5 have not yet been made. In their experiments so far, DeFelice *et al.* (1981), instead of trying to gather data at a wide range of frequencies, have concentrated on a particular one, which they call the "zero phase"

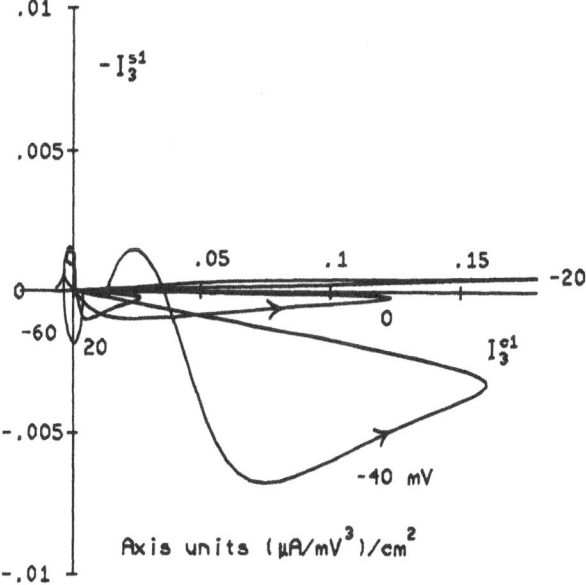

Figure 4. Third-order generalized admittances for the fundamental frequency.

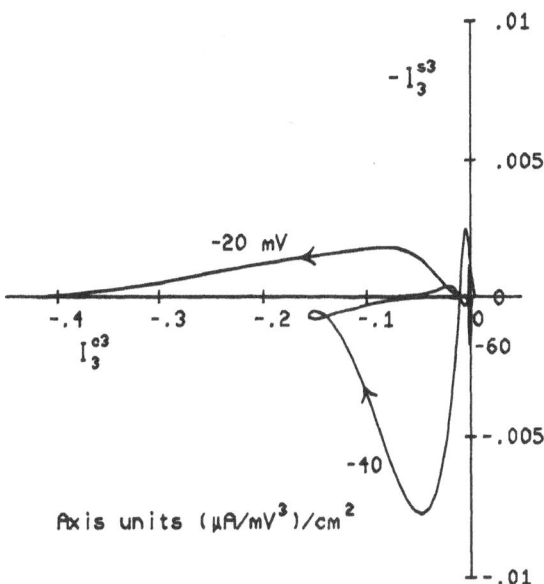

Figure 5. Third-order generalized admittances for three times the fundamental frequency.

frequency. That is the frequency at which the first-order admittance locus crosses the horizontal axis (Figure 1). Except for the curve at -40 mV, this crossing point lies to the right of the origin, and thus corresponds to a phase of zero for the admittance. For the -40 mV curve, however, this intersection lies to the left of the origin, and thus to a phase of 180° instead of 0°. (This crossing of the origin for a certain potential and frequency apparently did not occur in these experiments.) Thus it would be more precise to call this the "real admittance" frequency rather than the "zero phase" frequency. From the admittance loci it can also be seen that the amplitude of the linear component of current (the distance of a point on a curve to the origin) is approximately a minimum for this frequency f.

With V_0 and f fixed, the current waveform was recorded at different values of the input amplitude V_1, and separated into components at frequencies $0, f, 2f,$ and $3f$, using Fourier analysis. The amplitudes of the resulting current components were then plotted against V_1. With logarithmic scales on both axes, the result is a straight line for each frequency, with a slope which is equal to the order (1, 2, or 3). This result is easily explained by the theoretical analysis, since it is the natural result of plotting the different powers of V_1 which appear in equation (6) using logarithmic scales on both axes.

The theoretical lines resemble qualitatively those obtained from the experiments described by DeFelice *et al.* (1981) and also from other experiments done by Moore *et al.* (1980). A detailed comparison between theory and experiment must await completion of the analysis of the data. However, it appears so far that there is a reasonably good qualitative agreement between them. Further experiments and computations are needed.

REFERENCES

Chandler, W. K., FitzHugh, R., and Cole, K. S. (1962). Theoretical stability properties of a space-clamped axon, *Biophys. J.* **2**, 105–127.

Cole, K. S. (1968). *Membranes, Ions and Impulses* (University of California Press, Berkeley and Los Angeles).

DeFelice, L. J., Adelman, W. J., Jr., Clapham, D. E., and Mauro, A. (1980). Second order admittance in squid axon, *Fed. Proc.* **39** (6), Abstract No. 2465.

DeFelice, L. J., Adelman, W. J., Jr., Clapham, D. E., and Mauro, A. (1981). This volume, *The Biophysical Approach to Excitable Systems* (Plenum, New York), Chap. 3.

Moore, L. E., Fishman, H. M., and Poussart, D. J. M. (1980). Small-signal analysis of K^+ conduction in squid axons, *J. Membr. Biol.* **54**, 157–164.

Victor, J., and Shapley, R. (1980). A method of nonlinear analysis in the frequency domain, *Biophys. J.* **29**, 459–483.

Wiener, N. (1958). *Nonlinear Problems in Random Theory* (Technology Press of MIT and Wiley, New York).

3

Second-Order Admittance in Squid Axon

L. J. DeFELICE, W. J. ADELMAN, JR.,
D. E. CLAPHAM, AND A. MAURO

LINEAR AND NONLINEAR PROPERTIES OF NERVE MEMBRANE

The electrical behavior of nerve membranes can be divided into linear and nonlinear domains based on the response of the membrane to an external stimulus. Although the normal activity of the membrane is nonlinear, some properties of the membrane can be approximated by linear circuit elements if the external stimulus is small. This quasilinear behavior of membranes is described completely by the membrane's impedance.

L. J. DeFELICE AND D. E. CLAPHAM • Department of Anatomy, Emory University School of Medicine, Atlanta, Georgia 30322.
W. J. ADELMAN, JR. • Laboratory of Biophysics, IRP, NINCDS, National Institutes of Health at the Marine Biological Laboratory, Woods Hole, Massachusetts 02543.
A. MAURO • Rockefeller University, 1230 York Avenue, New York, New York 10021.

Impedance is defined most conveniently in the frequency domain. One property of the impedance, or its inverse the admittance, is that it is independent of the stimulus used to measure it. In contrast, the nonlinear behavior of membranes usually is defined by a set of differential equations in the time domain and is dependent on the stimulus used to measure it. One purpose of this paper is to show how the nonlinear properties of membranes can be studied in the frequency domain by methods familiar from linear analyses and to show how the nonlinear properties of membranes depend on stimulus amplitude at a given frequency.

We are interested in the linear behavior of nerve membranes because they do approximate some properties of the membrane. Membrane capacitance due to the dielectric properties of the lipid bilayer is linear under most conditions. In practice this means that the capacitance measured as a part of the total membrane impedance is approximately the same capacitance that acts for large stimuli or during normal activity such as the propagation of action potentials. Also, some states of the nerve membrane, such as the hyperpolarized state, are nearly linear for most purposes. A third example for which the linear approximation is often made has to do with the spontaneous fluctuations that occur in nerve membranes. These fluctuations, called membrane noise, are usually small; it is often assumed that a membrane impedance can be defined and that this impedance is the source of the fluctuations. For example, membrane impedance is used to relate membrane current noise to membrane voltage noise. Another purpose of our paper is to test the validity of this assumption and to determine the dependence of the definition of impedance on membrane voltage and frequency and lastly to compare our experimental results with the predictions of a standard model for nerve membrane. We shall show that under certain theoretical conditions the linear approximation breaks down completely and membrane impedance is undefined. We attempt to reproduce these conditions experimentally and compare our data with the theoretical treatment of nonlinearities in nerve membrane given by FitzHugh (1981) in this volume.

The techniques used to measure impedance and membrane noise are convenient and accurate and the results can be compared with specific models of the membrane to measure properties of interest, such as channel density and kinetics and single-channel conductance. These techniques and the results obtained from them are reviewed in DeFelice (1981). In this paper we wish only to make a connection between the linear and the nonlinear domains of nerve membranes and to show how the methods of linear system theory can be used to study the nonlinear behavior of nerve membranes. This is an alternative to the Wiener kernel analysis of nonlinear systems, first described by Wiener in 1958 and interpreted by Lee and Schetzen in 1965. As FitzHugh (1981) points out, the correspondence between theory and experiment using the present approach is incomplete and awaits further analysis. Here we concentrate on methods and present a limited set of experimental results for qualitative comparison with FitzHugh's treatment of the problem.

Although this paper deals with the properties of membranes under voltage clamp, the methods we describe are applicable to other problems. One example is the structural analysis of biological tissue using linear system theory. A general discussion of this field and a list of references are given in Eisenberg (1980), who points out that the structural analysis that relates impedance measurements to specific components of the tissue is not yet possible for the nonlinear response. Our attempt to treat the linear and nonlinear behavior of membranes by one method, and to quantify the nonlinearity and its dependence on the state of the system and the nature of the voltage perturbation, may find application in the measurement of the electrical parameters of tissue as well as the simpler case of isolated membranes.

Our results extend those reported by DeFelice *et al.* (1980) and Moore *et al.* (1980) [see Fishman *et al.* (1981), in this volume].

PROPERTIES OF LINEAR SYSTEMS

One fundamental property of any linear system is that its response to a sine wave perturbation is a sine wave at the same frequency. The response may be shifted in amplitude and phase compared to the forcing function, but if the system is linear the response will contain no frequency components other than the driving frequency. Furthermore, the amplitude of the response is related linearly to the forcing function; if the perturbation amplitude is doubled, the amplitude of the linear response also doubles.

Consider a nerve membrane under voltage clamp. If the membrane were linear, a sinusoidal voltage added to the steady membrane voltage would result in a pure sine wave current through the membrane at the driving frequency. The ratio of the current amplitude to the voltage amplitude defines the amplitude of the membrane admittance at each driving frequency. For a linear membrane, this ratio is independent of the amplitude of the sine wave. These criteria of linearity hold approximately for nerve membranes if the sine wave amplitude is small enough; however, this depends strongly on the perturbation frequency and on the average membrane potential.

THEORETICAL ADMITTANCE OF SQUID AXON MEMBRANE

Theoretically there is a membrane potential and a perturbation frequency for which no sine wave amplitude, however small, gives a linear response. This condition is illustrated in Figure 1. Figure 1 is derived from the linearized Hodgkin and Huxley (1952) equations for squid giant axon membrane. Small changes in membrane current, δI, are derived in terms of small changes in membrane voltage, δV. The ratio $Y = \delta I / \delta V$ is the membrane admittance. The details of the linearization are given by Hodgkin and Huxley (1952), Chandler *et al.* (1962), Mauro *et al.* (1970), and DeFelice (1981), and will not be repeated here. Once the admittance is defined it can be separated into its real and imaginary

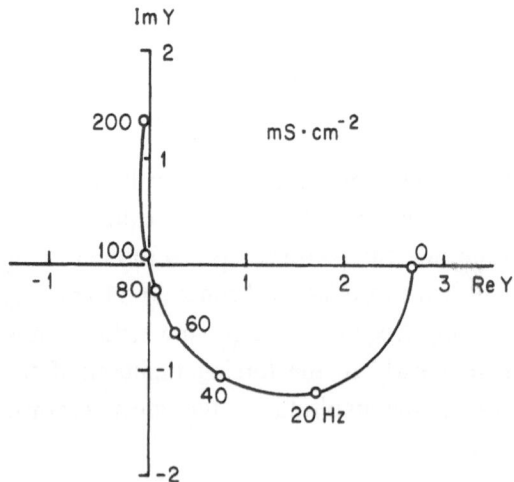

Figure 1. The theoretical admittance, $Y = a + ib$, of squid giant axon membrane in the complex plane; b is plotted against a for a 5.35-mV depolarization of the 1952 Hodgkin–Huxley equations at 6.3°C for frequencies between 0 and 200 Hz. The frequency at various points along the curve is indicated; the curve goes through the origin for $f = 93.3$ Hz.

parts; Figure 1 is a plot of the imaginary part against the real part for a specific membrane potential and temperature. At 0 Hz the imaginary component of the admittance is zero and the admittance is a point on the real axis. This is the dc conductance of the membrane. For any other frequency, the point lies in the complex plane and the locus of these points defines the total admittance of the membrane.

In our experiments we have used the peak-to-peak values of voltages and currents. In theoretical discussions we have employed FitzHugh's notation in which amplitudes (zero-to-peak values) of voltages and currents are used.

The membrane potential in Figure 1 was selected to cause the locus to go through the origin at some frequency. At any other voltage the locus crosses the real axis either to the right or the left of the origin, as shown in Figure 1 of FitzHugh (1981) in this volume.

Our Figure 1 says that for a particular membrane voltage and sine wave frequency, the total admittance of the membrane is zero, or the impedance is infinite. Theoretically, if a small voltage perturbation is applied under these conditions, no net current flows. This does not mean that no current flows through the membrane, only that the sum of the Na, K, and leakage currents in the Hodgkin–Huxley model is zero at these values of membrane potential and sine wave frequency.

Another approach to this phenomenon is to probe the nonlinear differential equations [the Hodgkin–Huxley (HH) equations] in the time domain with a small voltage sine wave added to the membrane potential. We are grateful to Dr. J. Cooley for providing us with such plots in the early stages of this study. He showed that under the conditions of Figure 1 for a 1 mV sine wave forcing function, the current at driving frequency vanishes a current at twice the driving frequency appears. This is true for arbitrarily small stimulus amplitudes. For the conditions of Figure 1 that cause the locus to pass through the origin, the theoretical nerve membrane is nonlinear at the origin regardless of the size of the sine wave.

EXPERIMENTAL TEST OF LINEARITY

To test this prediction we have clamped the squid giant axon with the potential

$$V(t) = V_0 + V_1 \cos(2\pi f t)$$

V_0 is the average membrane potential, i.e., the steady potential to which the membrane is clamped. The peak-to-peak amplitude (V_p) of the sinusoid added to the steady membrane potential is $2V_1$ and the stimulus frequency is f. FitzHugh (1981) has shown that the current response to this function is

$$I = I_0 + V_1 I_1 + V_1^2 I_2 + V_1^3 I_3 + \cdots$$

In FitzHugh's article, this expression for membrane current is written in the time domain. I_0 is the average membrane current in units of $\mu A/cm^2$. In the frequency domain, I_1 is identical to Y in this article and is the membrane admittance in units of millisiemens per square centimeter (mS/cm^2). I_2 is called the second-order admittance and has the units of $mS/mV\,cm^2$. Higher-order components of the admittance are defined similarly.

In this article we measure the amplitudes of the current response at the driving frequency and at $2f$, $3f$, etc. These amplitudes are measured in the frequency domain but are presented as the peak-to-peak values of the current components that would appear at these frequencies in the time domain. In FitzHugh's notation, the amplitudes of the first- and second-order components of membrane current are

$$|I_1| = \left[\left(I_1^{c1} \right)^2 + \left(I_1^{s1} \right)^2 \right]^{1/2}$$

$$|I_2| = \left[\left(I_2^{c2} \right)^2 + \left(I_2^{s2} \right)^2 \right]^{1/2}$$

Let f_r stand for the frequency for which the admittance locus of Figure 1 crosses the real axis. To the right of the origin, the current and the voltage are in phase at f_r. This is called the zero phase condition and occurs near the resonant frequency for a particular membrane potential. (This is the frequency for which the admittance is minimum, or the impedance is maximum.) To the left of the origin, the current and the voltage are 180° out of phase at f_r.

The ratio of $|I_2|$ to $|I_1|$ is a measure of the nonlinearity of the membrane. If the membrane were linear, all terms beyond the term $V_1 I_1$ in FitzHugh's equation for I would be zero. The ratio $|I_2|/|I_1|$ has the units mV^{-1}. If the ratio has the value $0.05/mV$, for example, this means that the amplitude of the second-order current, i.e., the current that appears at twice the sine wave frequency, is 5% of the current that appears at the forcing function frequency, for each millivolt amplitude of the sine wave: for $V_p = 1$ mV, the second-order current is 5% of the first-order current; but for $V_p = 2$ mV, the second-order current is 10% of the first-order current, and so on.

The final two graphs of this article compare our experimental results for this ratio with those predicted from the Hodgkin–Huxley model for the squid giant axon.

DATA ANALYSIS

Data were analyzed off-line on an HP-5420A digital analyzer. Three modes of analysis were used: for low-frequency forcing function sine waves (25 Hz), a bandwidth of 0–200 Hz was used with a frequency resolution of 781.25 mHz. For high-frequency sine waves (200 Hz), a bandwidth of 0–800 Hz was used with a frequency resolution of 3.125 Hz. For zero phase sine waves, a bandwidth of 0–400 Hz was used with a frequency resolution of 1.5625 Hz. In each case, 40 spectra were averaged for a total data collection time for the three cases of 51.2, 12.8, and 25.6 sec, respectively. The zero phase condition was determined on an oscilloscope during the experiments by adjusting the sine wave frequency. The true phase relationship between membrane current and membrane voltage was measured after the experiments. A positive phase (e.g., $\phi = 35°$, Figure 7) means that current lags voltage; a negative phase (e.g., $\phi = -73°$, Figure 7) means that current leads voltage. The sine wave amplitude (V_p) was adjusted with a potentiometer to nominal values between 100 μV and 16 mV but was measured separately for each experiment for the calculation of the admittance amplitudes and for the calculation of the ratio of second-order to first-order admittance amplitudes.

PREPARATION AND ELECTRONICS

Intact, cleaned single giant axons isolated from the medial stellar nerve of the squid, *Loligo pealei* were used. These axons were voltage clamped in a chamber located in a Faraday box made of irridited (chromium-plated) plate aluminum. The voltage clamp was a battery-operated low-noise version of the system described by Adelman and French (1978).

All the pertinent electronics (voltage control circuit, internal and external voltage amplifiers, and current amplifiers) were located inside the Faraday box. Manual controls of the voltage clamp system were operated by nonconducting shafts extending through waveguides in the walls of the Faraday box. Signal grounds were isolated from chassis or power grounds to reduce 60-Hz pickup.

Two voltage clamp commands were generated: (1) a dc voltage used to set the holding potential, and (2) a sine wave of variable frequency and voltage amplitude. Membrane voltage and current were recorded on a twin-trace cathode ray oscilloscope and on an FM tape recorder.

Temperature was controlled by a countercurrent heat exchanger using ice water flowing from an insulated reservoir contained within the Faraday box. The composition of the artificial seawater (ASW) used as the external medium was the same as that described by Adelman and French (1978). In some experiments ASW containing 1 μM tetrodotoxin (TTX-ASW) was used.

STANDARD MEASUREMENT OF MEMBRANE ADMITTANCE

The theoretical result of Figure 1 predicts that at some membrane potential V_0 and some perturbation frequency f, the real and the imaginary parts of the admittance of squid giant axon membrane are zero; this means that the amplitude of the admittance (the modulus) is zero at these values of V_0 and f. Figure 2 is an attempt to reproduce this condition experimentally. The axon was clamped to its resting potential of -60 mV. To measure the amplitude of the admittance, a 1-mV sinusoidal wave was added to the voltage clamp and the frequency of this wave was varied continuously between 20 and 200 Hz over a period of 30 sec. This is slow enough to assume nearly steady-state conditions at each frequency. The current through the membrane was measured with an rms meter with a 3-db lower cutoff frequency of 10 Hz. The current is divided by the rms of the sinusoidal stimulus and membrane area and is plotted in Figure 2 as membrane conductance per unit area as a function of stimulus frequency. The membrane was then clamped to five depolarizing poten-

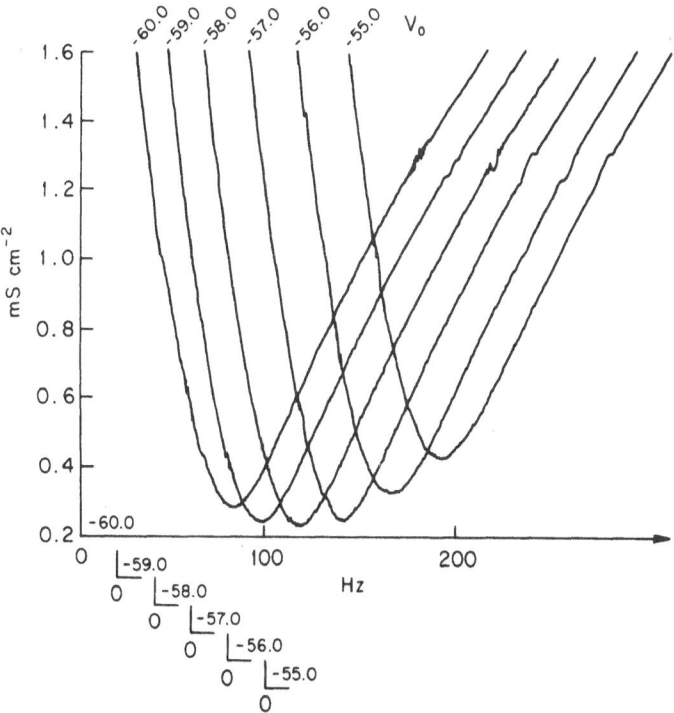

Figure 2. The measured admittance of the squid giant axon membrane in ASW at 7°C. The axon was clamped at its resting potential ($V_0 = -60$ mV) and depolarized in 1 mV steps to -55 mV. At each potential a 1 mV (peak-to-peak) sine wave (V_p) was added to the clamp. The frequency of V_p was swept slowly between 20 and 200 Hz. The figure is the output of an rms meter monitoring the steady-state sinusoidal component in current induced by V_p. The vertical axis is scaled to conductance per unit area. The horizontal axis is shifted to the right by 20 Hz for each value of V_0.

tials in 1-mV steps and the frequency was swept at each potential. To avoid overlap, the frequency axis in Figure 2 is shifted to the right by 20 Hz for each experiment.

The admittance goes through a minimum at each holding potential at some particular frequency. This is a well-known result and agrees qualitatively with the theoretical admittance predicted from the Hodgkin–Huxley equations (see, for example, Poussart *et al.*, 1977). The minimum value of all minima shown in Figure 2 occurs at -58 mV near 80 Hz. Although these exact values change from axon to axon, the pattern shown in Figure 2 is true generally; at no value of membrane potential or stimulus frequency, however carefully one attempts to ex-

plore these variables, does the amplitude of the admittance go to zero by
this method.

LINEARITY AND NONLINEARITY IN THE TIME DOMAIN

Figure 3 is the same type of experiment shown in Figure 2, with the
raw data viewed in the time domain before analysis. First consider
Figures 3a–3c. Only the 1 mV sinusoidal voltage perturbation (smooth

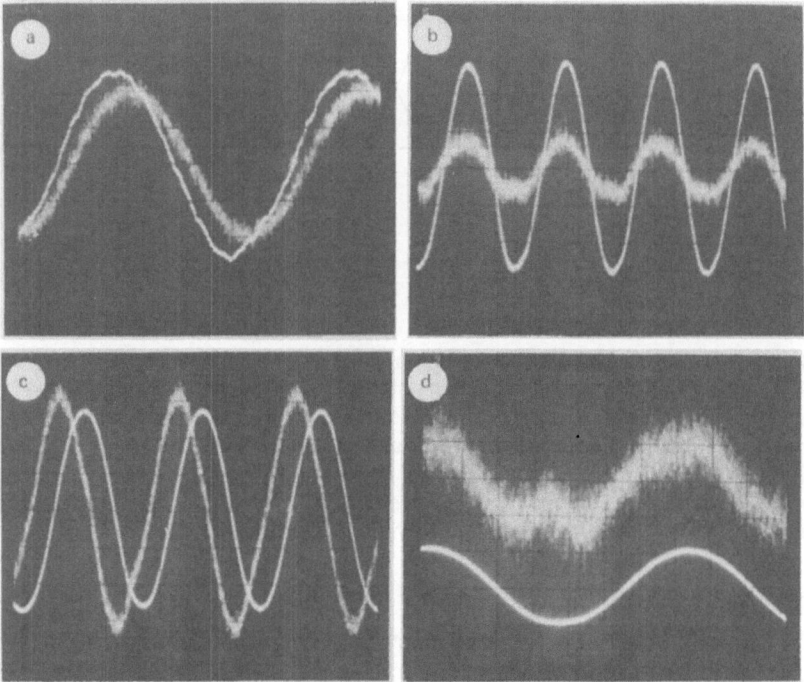

Figure 3. The admittance of the squid giant axon membrane in time domain. (a)–(c)
are from the same axon clamped at rest ($V_0 = -57.5$ mV) in ASW at 9°C. A 1 mV
peak-to-peak sine wave (smooth traces) was added to V_0. The fuzzy traces are
membrane current. Time scales are different for each frame: (a) 15 Hz, current lags
voltage. (b) 73 Hz, current and voltage at zero phase, (c) 300 Hz, current leads
voltage. 73 Hz is the frequency at which the minimum current is obtained. 100
nA/div, 0.24 cm^2. (d) Different axon at the absolute minimum current at any
frequency or voltage. $V_0 = -55.1$ mV, 71.4 Hz, 8°C; $V_p = 1$ mV; 20 nA/div, 0.24 cm^2
(from DeFelice, 1981).

trace) and the membrane current (fuzzy trace) are shown; both were ac coupled at 1 Hz so their average values are removed. The membrane was clamped to its resting potential of -57.5 mV. Figure 3b shows the result when a frequency is selected at which the minimum current amplitude occurs; at this frequency (73 Hz in this axon) the current is nearly in phase with the voltage. We call this the zero phase condition. [Figure 3b corresponds to the minimum admittance at the resting potential $(-60$ mV) shown in Figure 2.]

For a frequency on either side of the zero phase condition, the current amplitude increases: at lower frequencies, Figure 3a, current lags voltage; this is an inductive response. At higher frequencies, Figure 3c, current leads voltage; this is a capacitive response. (Figures 3a and 3c correspond to values on either side of the minimum in the -60 mV curve of Figure 2.)

In the current trace of Figure 3b, notice a slight flattening of the current in the troughs of the wave. This is shown in more detail in Figure 3d from a different axon depolarized by 3 mV. In this axon the zero phase frequency occurred at 71.4 Hz at -55.1 mV. This was the voltage for which the current amplitude appeared to be at its absolute minimum value at any frequency. (Figure 3d corresponds to the minimum in the -58 mV curve shown in Figure 2.) In addition to the first-order component of the current (the component that appears to be in phase with the voltage in Figure 3d) a second component at $2f$ is now obvious. For small values of the sine wave amplitude (V_p), such as the 1 mV sine wave used in Figures 2 and 3, the $2f$ component of the current adds to the first-order component in a characteristic way at the frequency for which the minimum occurs: near the zero phase frequency, the $2f$ component adds symmetrically to the troughs and to the peaks of the f component of current.

The measurement of the admittance is clearly more complex than indicated by the method of Figure 2. Near the minima, the membrane shows nonlinear behavior even for a relatively small perturbation amplitude. The current contains frequency components other than the driving frequency. This is most evident near the membrane potential and stimulus frequency where the first-order component of the current has its lowest value. However, at no value of membrane potential or perturbation frequency does the first-order component of the current observed in the time domain go to zero as predicted.

LINEARITY AND NONLINEARITY IN THE FREQUENCY DOMAIN

In order to determine the amplitudes of the various frequency components in the membrane current, data like those shown in Figure 3 were analyzed in the frequency domain. Figure 4 shows an example of such an analysis for one axon. The vertical axis is scaled to give the

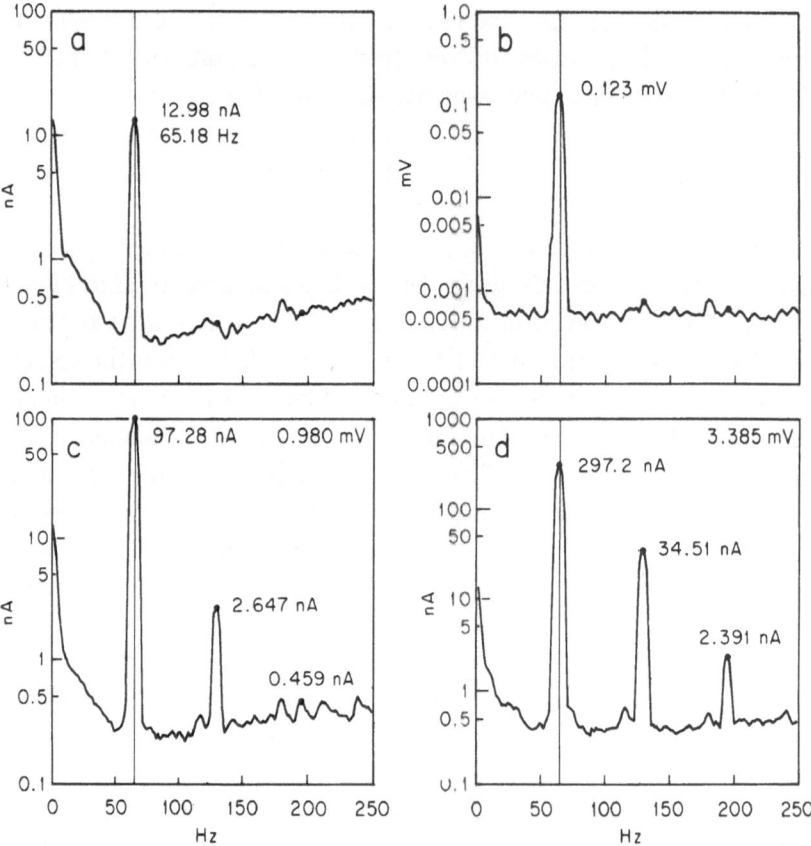

Figure 4. An example of frequency domain analysis of data like those shown in Figure 3. All frames are from the same axon clamped at rest ($V_0 = -61$ mV) in ASW at 7°C. The vertical axis is proportional to the modulus of the Fourier transform of current or voltage, scaled to give the peak-to-peak amplitude of the sinusoidal component of current or voltage at any frequency. (a) is the current response to (b), a voltage perturbation $V_p = 123$ μV at 65.18 Hz. This frequency corresponds to the zero phase condition ($\phi = -3.2°$). (c) and (d) show the current response to increase V_p at the same frequency. The amplitude of V_p is given in the upper right-hand corner. The dots in each frame occur at $f = 65.18$ Hz and at $2f$ and $3f$.

peak-to-peak value of the current components in the time domain. The membrane was clamped to its resting potential of -61 mV. A sine wave frequency was selected near the zero phase condition, which for this potential was 65.18 Hz. The analysis of the forcing function for $V_p = 123$ μV (Figure 4b) shows that to a good approximation the wave added to the clamp potential is a pure sinusoid. That is, there are no $2f$ or $3f$ components in the forcing function. The analysis of the current response to this perturbation is shown in Figure 4a. For 123 μV, only the first-order or linear component of the current is excited. For higher perturbation amplitudes, shown as a parameter in the upper right-hand corner of Figures 4c and 4d, second- and third-order current components appear. Increasing the sine wave amplitude from about 1 mV (Figure 4c) to about 3.4 mV (Figure 4d) more than triples the first-order current component, as would be expected; the second-order component is increased by over a factor of 10.

DEPENDENCE OF FREQUENCY COMPONENTS OF MEMBRANE CURRENT ON STIMULUS AMPLITUDE

Figure 5 shows how the amplitude of the first- and higher-order components of membrane current depend on the amplitude of the stimulus for both small and large values of V_p. The axon was depolarized by 10 mV and then stimulated by a sinusoid added to the clamp. The frequency of the forcing function was selected near the zero phase condition and the sine wave amplitude (V_p) was varied in steps between 125 μV and 16 mV. The data points are the peak-to-peak value of the current per unit area at f, $2f$, $3f$, and $4f$, where f is the perturbation frequency.

For values of V_p below 5 mV the first- and second-order currents are proportional to V_p and V_p^2. Above 5 mV the first-order currents lie below a line of slope one and the second-order currents lie above a line of slope two. The data points from the $3f$ component also tail upward compared to a line of slope three fitted to the points at low values of V_p. The $4f$ component was barely excited in this experiment; at values of V_p below 10 mV, the $4f$ component is proportional to V_p, suggesting that it is due not to membrane nonlinearities but to $4f$ components in the stimulus; above 10 mV, the $4f$ component is proportional to V_p^4 as shown.

The rather complex behavior shown in Figure 5 is simplified if the sine wave amplitude is kept below 3 mV and the average clamp potential is held near rest. Figure 6 shows an axon depolarized by 4 mV. The perturbation frequency is near the zero phase condition where the first-order current is minimum. The f and $2f$ components of the current are proportional to V_p and V_p^2 over this range of V_p to a good approximation. A slight droop in the first-order current is seen around 2 mV. This is a regular feature of such curves and will be discussed below.

DEPENDENCE OF FIRST- AND SECOND-ORDER CURRENTS ON STIMULUS FREQUENCY AND MEMBRANE VOLTAGE

Figures 4, 5, and 6 focus on the zero phase condition in which the sine wave frequency is selected near the minimum of the admittance. In

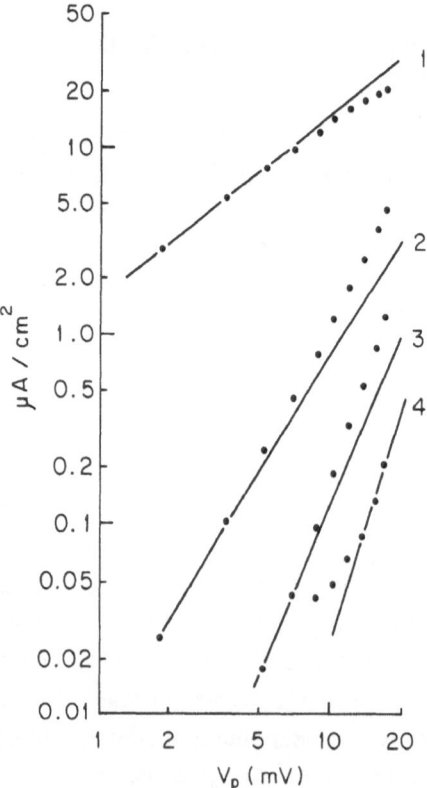

Figure 5. Squid axon membrane clamped to $V_0 = -49$ mV in ASW at 4.7°C. The resting potential was -59 mV. A 62.2-Hz sine wave with peak-to-peak amplitude V_p was added to V_0. This frequency corresponds to $\phi = -7.6°$ for $V_p = 125$ μV. The vertical axis is the peak-to-peak current per unit area measured at the driving frequency (1), and at twice (2), three times (3), and four times (4) this value. These currents are plotted as a function of V_p. Straight lines of slope 1, 2, 3, or 4 have been drawn through the data by eye.

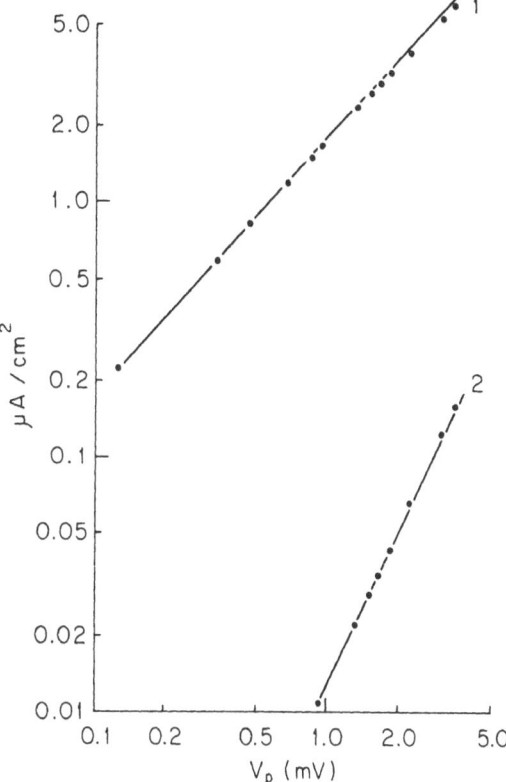

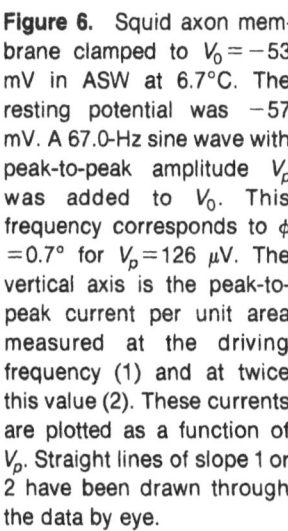

Figure 6. Squid axon membrane clamped to $V_0 = -53$ mV in ASW at 6.7°C. The resting potential was -57 mV. A 67.0-Hz sine wave with peak-to-peak amplitude V_p was added to V_0. This frequency corresponds to $\phi = 0.7°$ for $V_p = 126$ μV. The vertical axis is the peak-to-peak current per unit area measured at the driving frequency (1) and at twice this value (2). These currents are plotted as a function of V_p. Straight lines of slope 1 or 2 have been drawn through the data by eye.

Figure 7 we show the effect of frequency on the f and $2f$ components of current near rest. The stimulus frequencies are 67.7 Hz, the zero phase condition at -61 mV in this axon, and 200 Hz and 25 Hz on either side of the minimum admittance. The first-order component of current (the top three lines in Figure 7) agree qualitatively with the data shown in Figures 2 and 3. The analysis shows that the $2f$ component does not go through a minimum in this frequency range, but appears to fall off monotonically for f between 25 and 200 Hz. Values of f and $2f$ are shown as parameters near the graphs in Figure 7.

The ratio of the amplitude of the second-order admittance, $|I_2|$, to the amplitude of the first-order admittance, $|I_1|$, was calculated for each frequency in Figure 7. This was done by dividing the $2f$ component of current by the f component and by V_p at every value of V_p:

$$\frac{\left(\text{peak-to-peak value of the } 2f \text{ component of current at } V_p\right)}{V_p \times \left(\text{peak-to-peak value of the } f \text{ component of current at } V_p\right)} = \frac{|V_1{}^2 I_2|}{V_1 |V_1 I_1|} = \frac{|I_2|}{|I_1|}$$

These ratios are given in the legend of Figure 7. The ratio, which is a measure of the membrane's nonlinearity, is largest for the zero phase condition, i.e., at a frequency near the minimum admittance.

In Figure 8 we show the effect of membrane voltage on the first- and the second-order currents at the zero phase condition near rest. Hyperpolarizing the membrane decreases the first-order component as expected; the second-order component is also decreased. Depolarizing the membrane, however, increases the first-order component but decreases the second-order component. Notice the droop in the first-order curves at

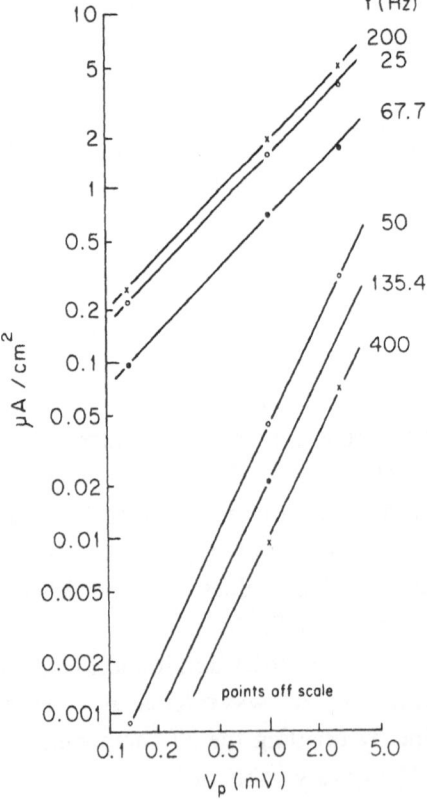

Figure 7. Squid giant axon in ASW at 5.6°C, clamped to its resting potential, $V_0 = -61$ mV. The top three lines show the current per unit area measured at the frequency of the sine wave added to the clamp potential (V_0). This frequency is shown by each curve. The bottom three lines show the current measured at twice the driving frequency. The first- and second-order currents are plotted as a function of V_p. When $f=67.7$ Hz, the minimum current at this potential was recorded ($\phi = -11.6°$). The frequencies 25 and 200 Hz were arbitrarily selected on either side of this minimum, and for these frequencies, $\phi = 35°$ and $-73°$. Straight lines of slopes one and two are drawn through the points by eye. The ratio of the amplitude of the second-order admittance to the first-order admittance was calculated for each frequency: For $f=25$ Hz, $|I_2|/|I_1| = 0.0269 \pm 0.0002$ mV^{-1}; for $f=67.7$ Hz, the ratio is 0.0331 ± 0.0022 mV^{-1}; for $f=200$ Hz, the ratio is 0.0046 ± 0.0001 mV^{-1} ($n = 3$).

Figure 8. Squid giant axon in ASW at 6.9°C, clamped to V_0 = −58 mV (the resting potential was −61 mV) and to two other potentials 5 mV on either side of −58 mV. In each case, a sine wave was added to V_0 at a frequency selected near the zero phase condition. The values of f and ϕ for three values of V_0 are 83.6 Hz, −3.4° at −53 mV, 56.0 Hz, −4.9° at −58 mV, 27.2 Hz, 0.9° at −63 mV. ϕ was measured at V_p =1 mV. The vertical axis is the peak-to-peak value of the current per unit area measured at the driving frequency and at twice the driving frequency. Straight lines of slopes one and two are drawn through the points by eye. For $V_0 = -53$ mV, the ratio of the amplitude of the second-order admittance to the first-order admittance is $|I_2|/|I_1| = 0.0150 \pm 0.0020$ mV^{-1}; for $V_0 = -58$ mV, the ratio is 0.0295 ± 0.0014 mV^{-1}; for $V_0 = -63$ mV, the ratio is 0.0227 ± 0.0005 mV^{-1} (calculated for $\frac{1}{2} < V_p < 2$ mV, $n=4$).

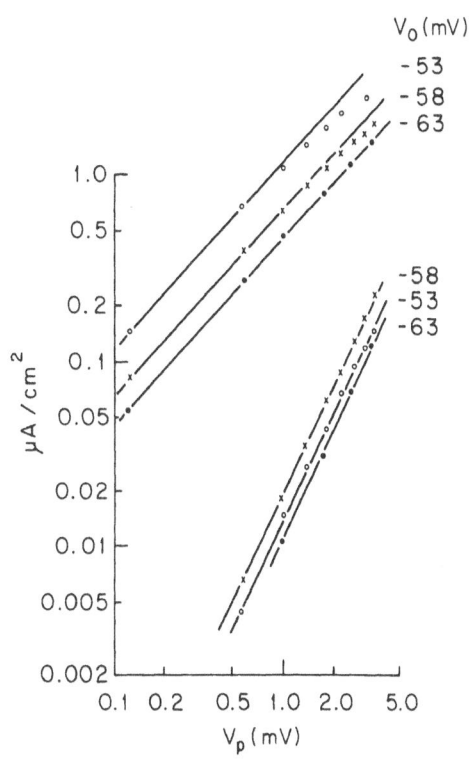

the higher values of V_p. The ratio $|I_2|/|I_1|$ is the largest in this axon for $V_0 = -58$ mV.

DEPENDENCE OF FIRST- AND SECOND-ORDER CURRENTS ON Na CURRENT

The effect of the Na current blocker TTX on the first- and second-order components of membrane current for the zero phase condition at the resting potential is shown in Figure 9. Blocking the Na current increases the first-order current because at the zero phase condition the Na current subtracts from other membrane currents to produce a minimum current. The effect of TTX on the second-order current is opposite; although TTX decreases the second-order current, the ratio $|I_2|/|I_1|$ is smaller when the Na current is eliminated.

Figure 10 shows the effect of varying V_p on the first- and second-order currents in the presence of TTX at two values of V_0 and two frequencies. In the presence of TTX the zero phase condition occurred at 48.7 Hz when V_0 was set equal to the resting potential (open circles). This is a lower frequency than normally obtained in ASW at rest for the zero phase condition. (Compare Figure 10, open circles, with Figure 7 at 67.7 Hz and Figure 9 at 58.4 Hz.) Decreasing the frequency to a value below the zero phase frequency increases both the first-order current and the second-order current for an axon at rest in TTX. For a 5-mV depolarization ($V_0 = -56$ mV, filled circles), the first-order current is increased at the zero phase frequency (74.7 Hz) at this potential. The first-order current also is increased at a lower frequency (25 Hz). In the depolarized state, the $2f$ component of the current at 25 Hz stimulation (labeled 50 in the graph) is larger than the $2f$ component for the zero phase frequency. This is the same effect observed in ASW in Figure 7. However, the second-order current for both the resting state and the depolarized state

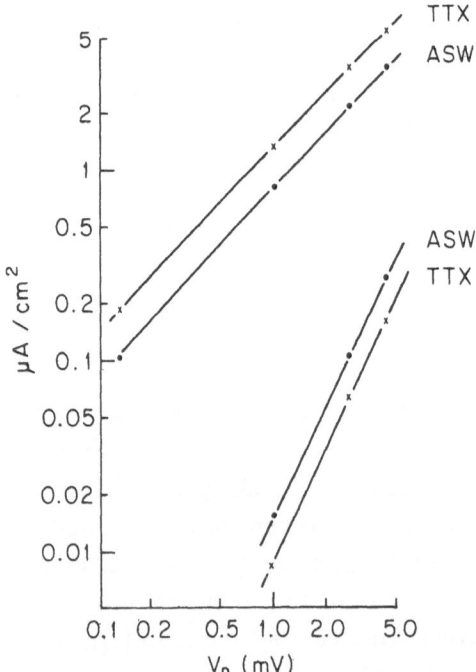

Figure 9. Squid giant axon membrane clamped to its resting potential $V_0 = -59$ mV at 6.4°C. A sine wave of value V_p was added to V_0 at a frequency 58.4 Hz, near the zero phase condition ($\phi = 3.0°$ measured at $V_p = 1$ mV) in ASW. The external solution was then changed to TTX-ASW and the experiment was repeated under the same conditions. (In TTX-ASW, $\phi = -0.7°$ measured at $V_p = 1$ mV at $f = 58.4$ Hz). The vertical axis is the peak-to-peak value of the current per unit area measured at the driving frequency and at twice the driving frequency. Lines of slopes one and two are drawn through the data by eye. In ASW, the ratio of the amplitude of the second-order admittance is $|I_2|/|I_1| = 0.0161 \pm 0.0022$ mV^{-1}; in TTX-ASW, the ratio is 0.0064 ± 0.0001 mV^{-1} ($n=3$).

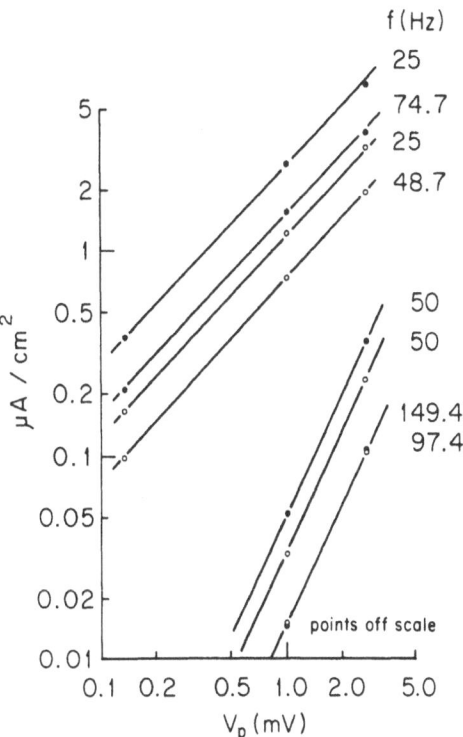

Figure 10. Squid giant axon membrane clamped to its resting potential $V_0 = -61$ mV and to -56 mV, in TTX-ASW at 5.8°C. A sine wave of value V_p was added to V_0. Two frequencies were used for each potential: For -61 mV, $f=48.7$ Hz, $\phi = -0.7°$; $f=25$ Hz, $\phi=2.8°$. For -56 mV, $f=74.7$ Hz, $\phi = -0.7°$; $f=25$ Hz, $\phi=27.8°$. In each case, ϕ was measured at $V_p=1$ mV. The vertical axis is the current amplitude measured at the driving frequency and at twice the driving frequency. Lines of slopes one and two are drawn through the data by eye.

nearly superimpose for the two different zero phase frequencies at the two values of V_0. To summarize Figure 10, the presence of TTX, depolarizing the membrane by 5 mV increases the frequency at which the zero phase condition occurs, and increases the f component of current if the two are compared at their respective zero phase frequencies. At 5-mV depolarization, the first- and second-order currents are nearly equal at their respective zero phase frequencies.

EXPERIMENTAL AND THEORETICAL NONLINEARITY COMPARED

This paper is concerned primarily with the nonlinearity of squid axon membrane under various conditions. We use the ratio of the amplitudes of the second-order admittance to the first-order admittance as an index of nonlinearity. This is convenient experimentally, for it is easily measured from the current amplitudes (see the discussion near Figure 7) and allows membranes of different areas to be compared since

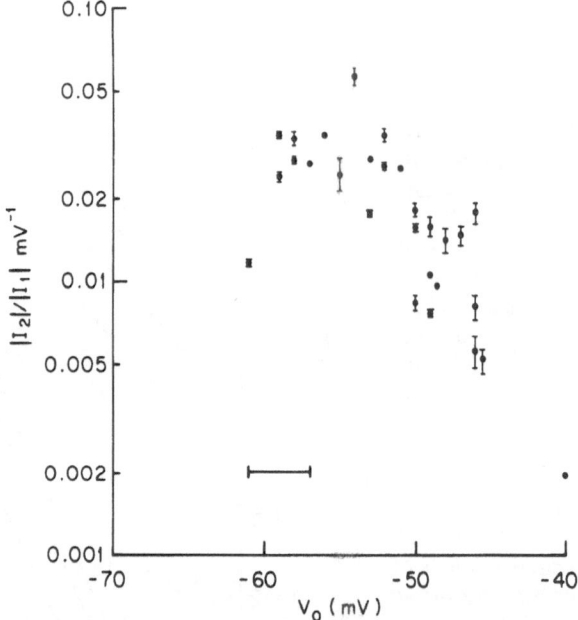

Figure 11. Pooled data from 10 axons in ASW. The vertical axis is the ratio of the amplitude of the second-order admittance $|I_2|$ to the linear, first-order admittance $|I_1|$. This ratio is plotted as a function of the mean clamp potential V_0. All data are taken with $1 < V_p < 5$ mV, $-8° < \phi < 8°$, $4.7°C < T < 7.1°C$. The zero phase frequencies at different values of V_0 range from 27 to 89 Hz. The error bars are standard deviations for the ratio $|I_2|/|I_1|$ calculated for the same membrane under identical conditions but for n values of V_p, $2 < n < 6$. No error bars indicate $n=1$. $|I_2|/|I_1|$ was measured by dividing the peak-to-peak current amplitude at $2f$ by the peak-to-peak current amplitude at f and by V_p. The horizontal bar represents the range of resting potentials of the ten axons used in this plot.

the ratio is independent of area. Figure 11 plots this ratio as a function of membrane voltage for the zero phase frequency for each voltage. The zero phase condition was selected because here the theoretical admittance is near its minimum value; we assumed that second-order currents would be largest, relative to first-order currents, when the first-order currents have their smallest value. Since the first-order current is theoretically zero at some potential, we measure the ratio as a function of membrane potential to see where the membrane is most nonlinear by our criterion.

Figure 11 shows that the largest second-order currents, relative to first-order currents, occur for 2–6-mV depolarizations. The largest value of this ratio found in our experiments under normal conditions was 0.058/mV at -54.3 mV for an axon whose resting potential was -59.5 mV. For this axon, then, the second-order current was about 6% of the first-order current for a 5-mV depolarization for each millivolt of stimulus amplitude at the zero phase frequency. For a sine wave amplitude ± 2 mV around 54 mV, the second-order current would be one-fourth the size of the first-order current.

The scatter in Figure 11 is due partly to the spread of experimental conditions in the ten axons used. The resting potential of these axons varied between -61 and -57 mV. The zero phase condition was obtained by eye on the oscilloscope and was measured after each experiment to be only within $\pm 8°$ of zero phase. The temperature was constant to within $2.4°C$, but some values of V_p used in this analysis were as large as 5 mV, where other effects, to be discussed below, are present. Nevertheless, it is clear that the largest nonlinearities, by our criterion, occur at several millivolts depolarization from rest, and that depolarizing the membrane by about 10 mV decreases this nonlinearity by about a factor of 10 from its maximum value.

Figure 12 is the ratio $|I_2|/|I_1|$ as a function of membrane voltage calculated by R. FitzHugh from the 1952 Hodgkin–Huxley equations.

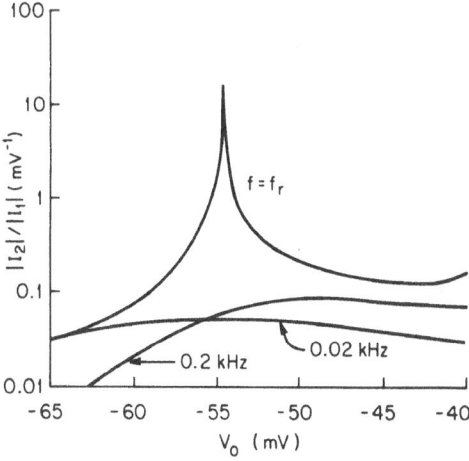

Figure 12. $|I_2|/|I_1|$ versus membrane potential, calculated from the 1952 Hodgkin–Huxley equations by R. FitzHugh (personal communication). The two lower curves are calculated for constant frequency, one at 20 Hz and the other at 200 Hz. The top curve is calculated at the frequency for which the imaginary part of I_1 is zero at each potential. This corresponds approximately to the zero phase frequency used in Figure 11; the theoretical plot also includes frequencies for which the admittance locus crosses the real axis to the left of the origin, where current and voltage are 180° out of phase. The curves stop at -65 mV (FitzHugh set the HH potential variable so that the resting potential is at -60 mV) because below this potential the admittance locus does not turn downward in the complex plane (see Figure 1) and hence does not cross the real axis, leaving f_r undefined.

The ratio is shown for 20 and 200 Hz in the bottom two curves. The top curve was calculated as follows: In FitzHugh's notation, I_1 is expressed as [FitzHugh, 1981, equation (7)]

$$I_1 = I_1^{c1}\cos 2\pi ft + I_1^{s1}\sin 2\pi ft$$

where I_1^{c1} and $-I_1^{s1}$ are the real and the imaginary parts of the admittance. The frequencies used in the top curve in Figure 12 are the nonzero positive roots of the equation

$$I_1^{s1} = 0$$

at each membrane potential, where I_1^{s1} is considered as a function of frequency. The frequency obtained is the frequency for which the admittance locus crosses the real axis for a given membrane potential. For some potentials (see FitzHugh, 1981, Figure 1) the theoretical admittance locus crosses the real axis to the left of the origin and current and voltage are 180° out of phase. This was never observed experimentally. The experimental points in Figure 11, to which the upper curve in Figure 12 should be compared, include only the zero phase conditions, although Figure 12 itself includes both the zero phase and 180° phase conditions depending on the potential.

In Figure 12, the ratio $|I_2|/|I_1|$ goes to infinity for V_0 between −54.6 and −54.7 mV. This corresponds to potential used in Figure 1 that causes the admittance locus to pass through the origin near 93 Hz. Since both the real and imaginary parts of the admittance are zero for this condition, $|I_1|$ is zero and $|I_2|/|I_1|$ goes to infinity. This was never observed experimentally. The experimental points shown in Figure 11 lie below the $f = f_r$ theoretical curve in Figure 12 at any value of V_0, because the measured admittance at the zero phase frequency never approaches zero.

Figure 12 does support the assumption, discussed in Figure 11, that the ratio $|I_2|/|I_1|$ is largest near the zero phase frequency, at least when compared to frequencies well below (20 Hz) and well above (200 Hz) the zero phase condition.

DISCREPANCIES BETWEEN THEORY AND EXPERIMENT

There are two major discrepancies between our results on membrane admittance and the theoretical admittance predicted by the Hodgkin–Huxley equations. We have never observed the theoretical state in which the first-order component of current is zero. This state corresponds to the point in Figure 1 where the admittance locus passes through the origin and the point in Figure 12 where $|I_2|/|I_1|$ goes to infinity. And we have never found a condition where current and voltage are 180° out of phase. Theoretically, this occurs between $V_0 = -54$ mV, at 93 Hz, and $V_0 = -38$ mV at 170 Hz. For both of these conditions, the admittance is zero; between these voltages the admittance locus crosses the real axis to the left of the origin and the current and voltage are exactly out of phase. Although we measured the admittance between -70 and -40 mV, at frequencies between 20 and 250 Hz, we never observed either the zero admittance state or the 180° phase shift.

There are several explanations for these discrepancies. First, the equations used in theoretical treatment do not apply exactly to our axons. Equations revised to correspond more closely to our nerves, such as those developed by Adelman and FitzHugh (1975) to account for K accumulation, may give results closer to our experiments. We have no evidence for or against this possibility. Second, experimental conditions may have resulted in our failure to observe predicted behavior. In our experiments, membrane potentials were held for approximately 60 sec while admittance was measured, and this relatively long clamp may have resulted in a deterioration of the axons during the experiments.

Fishman *et al.* (1981) show that in their experiments the admittance is the same whether data are collected over 1 sec or 20–40 msec and conclude that K kinetics are unchanged in either case. There is a drop in K current during long clamp periods, but this does not affect membrane admittance except at large depolarizations. Clamp periods comparable to ours were not studied.

Repeating our experiments with shorter clamp periods may remove the discrepancies mentioned above; however, we have at present no evidence for or against this. A third possibility is that Na current in our experiments was less than normal. Action potentials were typically 100 to 120 mV before an experiment but 80 to 120 mV after an experiment.

Clapham (1979) has shown that removing all Na current from the HH equations does eliminate the zero admittance condition; however, the effect of a partially reduced Na current is unknown.

Such explanations require further analysis. Here we discuss nonlinear phenomena that affect the interpretation of impedance measurements even in the absence of these experimental difficulties.

We have shown that nonlinearities in the membrane are a sensitive function not only of membrane voltage but also of stimulus frequency, and that nonlinearities may be appreciable for stimulus amplitudes normally considered small. Figure 1 says that there is a perturbing voltage and frequency for which there is no net current. This, however, is a prediction from the linearized Hodgkin–Huxley equations. Admittance measured by the method of Figure 2 hides nonlinear effects because higher-order frequency components are integrated with the first-order component. Although the membrane voltage perturbation amplitude is only 1 mV, these higher-order components stand out near the resonant frequency where the first-order component is smallest. This is shown in Figure 3. Thus, even if the current at the driving frequency were zero, nonlinear effects near the resonant frequency would give net currents at $2f$, $3f$, etc.

The method of Figure 2 would fail to give zero admittance in any case because of nonlinear effects, but the frequency domain analysis shown in Figure 4 also fails. We found no condition where the component of membrane current at the perturbing frequency was zero or where it was 180° out of phase with voltage. Instead, the f component of current goes through a nonzero minimum at some frequency where it is nearly in phase with voltage.

THE FREQUENCY COMPONENTS OF THE HIGHER-ORDER CURRENTS

Nonlinear effects also contribute to the membrane current at the driving frequency. In FitzHugh's notation, Chapter 2 of this volume, the second-order current is given by

$$V_1{}^2 I_2 = V_1{}^2 I_2^{c0} + V_1{}^2 \left(I_2^{c2} \cos 4\pi ft + I_2^{s2} \sin 4\pi ft \right)$$

The first term is a component of the second-order current at $f=0$. The second term is a component of the second order current at twice the forcing function frequency ($2f$). I_2^{c0} is plotted in FitzHugh's Figure 2 and I_2^{c2} versus I_2^{s2} is plotted in his Figure 3.

The third-order current is given by

$$V_1{}^3 I_3 = V_1{}^3 \left(I_3^{c1} \cos 2\pi ft + I_3^{s1} \sin 2\pi ft \right) + V_1{}^3 \left(I_3^{c3} \cos 6\pi ft + I_3^{s3} \sin 6\pi ft \right)$$

The first term is a component of third-order current at the stimulus frequency (f). The second term is a component of the third order current at three times the stimulus frequency ($3f$). I_3^{c1} versus I_3^{s1} is plotted in FitzHugh's Figure 4 and I_3^{c3} versus I_3^{s3} is plotted in his Figure 5.

In general, the even nonlinear terms contribute even harmonics up to their order, and the odd nonlinear terms contribute odd harmonics up to their order. The total current

$$I_0 + V_1 I_1 + V_1{}^2 I_2 + V_1{}^3 I_3 + V_1{}^4 I_4 + \cdots$$

has components (term by term) at

$$0 \quad f \quad 0,2f \quad f,3f \quad 0,2f,4f, \text{ etc.}$$

Unpublished calculations provided by FitzHugh show that at -60 mV and 54 Hz (zero phase) the current measured at the perturbation frequency falls below the expected line when plotted against the sine wave amplitude. This is because the f component of $V_1{}^3 I_3$ subtracts from the f component of $V_1 I_1$. Theoretically, the effect becomes appreciable for V_1 above 3 mV ($V_p > 6$ mV).

This may explain the droop seen in our first-order currents at higher values of V_p (see Figures 5–8). The f component of the first-order current is proportional to V_p but the f component of the third-order current is proportional to V_p^3. If the current measured at f is due to a subtraction of these components we expect it to fall increasingly below a line of slope 1 as V_p increases. This is a measurable effect in all our experiments for V_0 near rest and f near the zero phase condition.

In Figure 5, where V_p is largest, similar effects are seen for currents measured at $2f$ and $3f$ except that they now lie above the expected line. Consider the curve labeled 2; if the $2f$ component of $V_1{}^4 I_4$ adds to the $2f$ component of $V_1{}^2 I_2$ this could explain the upward tail in the data. No calculations have been done above the $V_1{}^3 I_3$ term, so we have no direct evidence that this is the case. To avoid such effects, all of the experiments beyond Figure 5 were done for V_p less than 3 mV. In this range, the currents measured at f and at $2f$ are proportional to V_p and V_p^2 to a good approximation and are therefore from the first- and second-order terms only.

SUMMARY

Unpublished calculations by FitzHugh show that the dependence of the second-order current, $V_1{}^2 I_2$, on frequency and at the membrane's resting potential is qualitatively similar to our data in Figure 7. At $V_0 = -60$ mV, $V_1 = 1$ mV the HH equations predict that the amplitude of the second-order current falls from about 0.045 $\mu A/cm^2$ to 0.031 $\mu A/cm^2$ between 10 and 100 Hz. Our experimental values are lower than these, but the same behavior is observed. At other membrane potentials, the dependence of the second-order current on frequency may not be monotonic; e.g., at $V_0 = -20$ mV, the theoretical calculations predict that the second-order current goes through a maximum near 50 Hz. We have measured the dependence of $V_1{}^2 I_2$ on frequency over a wide range of potentials, but these data will be presented elsewhere. Here we remark only that near the resting potential the membrane is most nonlinear, by our $|I_2|/|I_1|$ ratio, near the zero phase frequency. By the same criterion, removing the Na current decreases the nonlinearity as shown in Figures 9 and 10.

Our data on the nonlinearity of the membrane and its dependence on membrane voltage are summarized in Figure 11. The conclusion to be drawn from these experiments is this; although the experimental membrane is most nonlinear for the zero phase condition and for 2–6 mV depolarization, it is less nonlinear than the theoretical membrane shown in Figure 12. The singularity in the model is never observed and the measured values of $|I_2|/|I_1|$ at zero phase lie below the predicted values over the potential range studied. Our data provide a quantitative picture of the dependence of the second-order admittance in squid axon on membrane voltage and stimulus frequency and amplitude, but the quantitative comparison of our data with FitzHugh's theory, and the explanation of the discrepancies between our data and predicted results, await further analysis.

ACKNOWLEDGMENTS

We wish to thank Mr. Grady Kelly for technical assistance. L. J. DeFelice was supported, in part, by an Interagency Personnel Act Fellowship.

REFERENCES

Adelman, W. J., Jr., and FitzHugh, R. (1975). Solutions of Hodgkin–Huxley equations modified for potassium accumulation in a periaxonal space, *Fed. Proc.* **34**, 1323–1329.

Adelman, W. J., Jr., and French, R. (1978). Blocking of the squid axon potassium channel by external caesium ions, *J. Physiol.* **276**, 13–25.

Chandler, W. K., FitzHugh, R., and Cole, K. S. (1962). Theoretical stability properties of a space-clamped axon, *Biophys. J.* **2**, 105–127.

Clapham, D. E. (1979). A whole tissue model of heart cell aggregates: Electrical coupling between cells, membrane impedance, and the extracellular space, Ph.D. thesis, Emory University.

DeFelice, L. J., Adelman, W. J., Jr., Clapham, D. E., and Mauro, A. (1980). Second-order admittance in squid axon, 1980 ASBC/BS meeting.

DeFelice, L. J. (1981). *Introduction to Membrane Noise* (Plenum, New York). N.Y.

Eisenberg, R. S. (1980). Impedance measurement of the electrical structure of skeletal muscle, in *Handbook of Physiology*, L. O. Peachey and R. Adrian, Eds., American Physiological Society Series (Williams and Wilkins Publishing Co., Baltimore, Maryland) (to be published 1982).

Fishman, H. M., Moore, L. E., and Poussart, D. (1981). Squid axon K conduction: Admittance and noise during short versus long duration step clamps, this volume, *The Biophysical Approach to Excitable Systems* (Plenum, New York), Chap. 4.

FitzHugh, R. (1981). Nonlinear sinusoidal currents in the Hodgkin–Huxley model, this volume, *The Biophysical Approach to Excitable Systems* (Plenum, New York), Chap. 2.

Hodgkin, A. L., and Huxley, A. F. (1952). A quantitative description of membrane current and its application to conduction and excitation in nerve, *J. Physiol.* **117**, 500–544.

Lee, Y. W., and Schetzen, M. (1965). Measurement of the Wiener kernel of a nonlinear system by cross-correlation, *Int. J. Control* **2**, 237–254.

Mauro, A., Conti, F., Dodge, F., and Schor, R. (1970). Subthreshold behavior and phenomenological impedance of squid giant axon, *J. Gen. Physiol.* **55**, 496–523.

Moore, L. E., Fishman, H. M., and Poussart, D. J. M. (1980). Small-signal analysis of K^+ conduction in squid axons, *J. Membr. Biol.* **54**, 157–164.

Poussart, D., Moore, L. E., and Fishman, H. M. (1977). Ion movements and kinetics in squid axon. I. Complex admittance. *Ann. N.Y. Acad. Sci.* **303**, 355–379.

Wiener, N. (1958). *Nonlinear Problems in Random Theory* (MIT Press, Cambridge, Massachusetts).

4

Squid Axon K Conduction: Admittance and Noise during Short-versus Long-Duration Step Clamps

H. M. FISHMAN, L. E. MOORE, and D. POUSSART

INDUCTIVE REACTANCE IN THE IMPEDANCE OF SQUID AXON: THE FREQUENCY DOMAIN MANIFESTATION OF LINEARIZED ION-CONDUCTION KINETICS

About forty years ago K. S. Cole and R. F. Baker (1941) measured the longitudinal impedance of a squid axon between distant external electrodes. Figure 1a shows 12 impedance points that were determined with an ac bridge for frequencies spanning the range from 30 Hz to 50 kHz, and the extrapolated impedance locus through the data points in the complex plane. Since the electrodes were entirely outside of the axon, the cable impedance was removed by squaring the vector magnitude and

H. M. FISHMAN and L. E. MOORE • Department of Physiology and Biophysics, University of Texas Medical Branch, Galveston, Texas.
D. POUSSART • Department of Electrical Engineering, Université Laval, Quebec, Canada.

doubling its phase angle at each frequency point in Figure 1a to obtain the corresponding membrane impedance points and locus in Figure 1b (Cole, 1972). The departure of the locus at low frequencies (<500 Hz) from the characteristic semicircular arc of a parallel *RC* circuit and the crossing of the real axis to give a spiral curve indicated for the first time an inductive reactance. This positive reactance caused considerable consternation amongst biologists, who interpreted this measurement literally as implying a magnetic-field-dependent process in the membrane. It is now clear that this interpretation is not correct and that the low-frequency inductive behavior is the frequency-domain manifestation of time-and voltage-dependent ion conduction (Cole, 1947; Hodgkin and Huxley, 1952; Chandler, FitzHugh, and Cole, 1962). With the subsequent development of the potential control ("voltage clamp") concept by Cole (1949) the time course of ion conduction could be observed directly. After an initial capacitive transient, following a step change in membrane potential, the current response of membrane conductance was evident. With the successes of potential control, the low-frequency impedance was, for a time, almost totally ignored as an alternative and useful means of describing the linear properties of membrane ion conduction.

ADVANCES IN THE SPEED AND RESOLUTION OF IMPEDANCE OR ADMITTANCE MEASUREMENTS

Six years ago we were motivated by complexities in the interpretation of spontaneous ion-conduction fluctuation data (Fishman *et al.*, 1975) to develop a perturbation method that could aid in the interpretation of spontaneous fluctuation measurements. Since spontaneous electrical fluctuations in membranes fall into the category of small-amplitude random noise, we felt that the best procedure would be to use noise as a perturbation signal to measure the membrane impedance or its reciprocal, the admittance during step voltage clamps (Fishman, 1975; Fishman *et al.*, 1977b; Poussart *et al.*, 1977). This decision was based on the following important considerations: The Cole and Baker measurements were done entirely outside the axon, and required corrections for the cable impedance. With longitudinal, transmembrane electrodes (two for potential and two for current) the membrane admittance could be

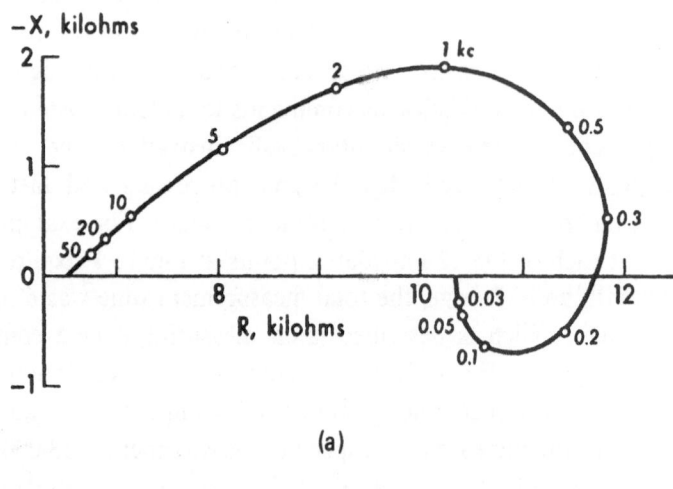

(a)

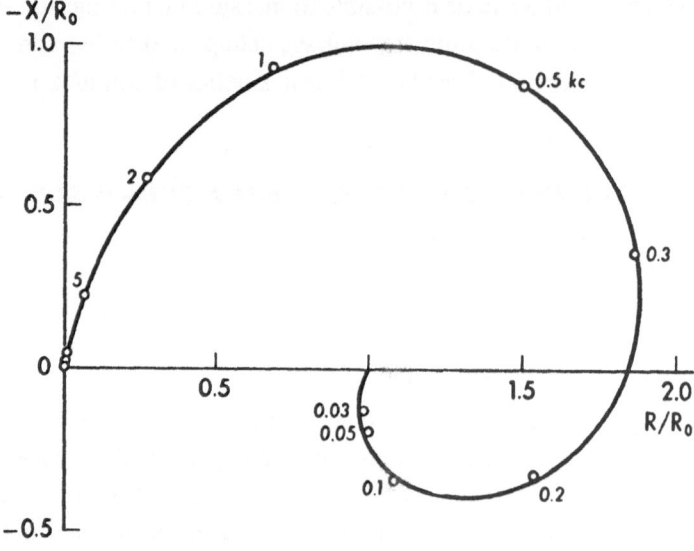

(b)

Figure 1. (a) Longitudinal impedance of a squid axon, measured by Cole and Baker (1941), between two external distant electrodes. (b) The membrane impedance after removal of the cable impedance in (a). After Cole (1972), by permission of University of California Press.

measured directly under potential control and with good signal-to-noise ratio under reasonably isopotential ("space clamp") conditions. A significant enhancement in the speed and detail of the measurement could be made by using synthesized, noise-like (pseudorandom) signals, with known spectral characteristics, as commands to a clamp system to apply many different frequencies simultaneously, instead of one at a time, together with synchronized, digital signal processing and fast Fourier transform computation of the membrane response. For example, if we assume that each of the 12 impedance points in Figure 1 required 30 sec to obtain with an ac bridge, the total measurement time was 6 min. The present system, which is described later, measuring over a comparable frequency range (50 Hz to 20 kHz), requires a data collection time of only 20 msec to produce a 400-point complex impedance or admittance function. Thus the increase in the speed of measurement is $18,000\times$, with an increase in resolution of $33\times$. This tremendous decrease in the measurement time has made it possible to measure membrane admittance during short- or long-duration step voltage clamps and to use admittance in new ways to describe directly the linear kinetics of conduction.

POTASSIUM CONDUCTION KINETICS FROM A COMPARISON OF ADMITTANCE AND NOISE DATA

In keeping with the theme of this volume, the data we present in this paper come from two types of measurement. In the first type, we have used two of Kacy Cole's major contributions—impedance determinations and the voltage clamp—in combination. The admittance of squid axon is measured for the potassium conduction process during membrane potential control for step changes with superposed small amplitude perturbations. The admittance provides a description of the linear conduction kinetics of the potassium system at various times after a step change in membrane potential. These data indicate that the potassium conduction system is not significantly altered in its kinetic properties for long-duration steps (sec) as compared to short-duration steps (msec). The second type of data presented in this paper is that of potassium current fluctuations. These also show that the power density spectrum, measured in the same axons in which the admittance is obtained, is not significantly altered for potassium current fluctuations during long-duration

step changes when compared to spectra of fluctuations during short-duration pulses. In other words, the potassium conductance kinetics obtained from steady-state analysis of either fluctuations or admittance in the squid axon exhibit the same properties as those derived during the characteristic time (msec) of potassium conduction. In addition, a qualitative comparison of the admittance with the power spectrum of potassium current fluctuations in the same axon yields essentially the same characteristic times for the two types of data at the same membrane potential. This agreement, over a limited (10-mV) potential range, and at three different temperatures between the two types of measurement, indicates that potassium conduction may be a linear kinetic process at the amplitude level of spontaneous fluctuations, since the admittance gives a linear description of conduction.

The evidence for the above statements is presented after a general discussion of linear ion conduction analysis via an admittance measurement, and a discussion of our implementation of this measurement.

STEADY-STATE LINEAR ANALYSIS OF CONDUCTION VIA ADMITTANCE

As a basis for the subsequent discussion of data, a brief, simplified description of admittance as it pertains to membrane ion conduction phenomena is given here. In Figure 2a, a linear, time-invariant conductance element, g, defines the relationship between the current through the element and the voltage across it. Substitution of a network of interconnected linear, time-invariant conductance elements as well as energy storage elements—capacitors and inductors, for g, in Figure 2b yields an analogous relationship between current and voltage, which defines the admittance—a complex quantity in the frequency domain.

In order to illustrate the connection between admittance and time- and voltage-dependent ion conduction, the temporal behavior of a series gL circuit (Figure 2c) is considered. The differential equation in the time domain, which relates the voltage, $v(t)$, across the gL circuit to the current $i(t)$ through it, is

$$v(t) = \frac{1}{g}i(t) + L\frac{di(t)}{dt} \tag{1}$$

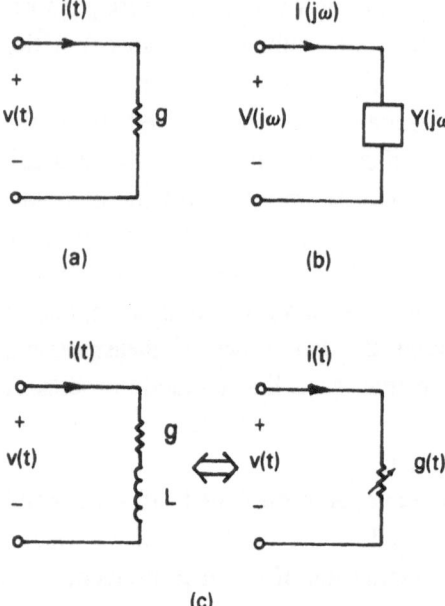

Figure 2. (a) A linear, time-invariant conductance, g. (b) An interconnected network of conductances and passive energy storage elements — capacitors and inductors. $Y(j\omega)$ is the admittance of the network. (c) A series gL circuit can exhibit the same kinetic behavior as a linear time-variant conductance (see text).

Equation (1) can be rewritten as

$$\frac{di(t)}{dt} = \frac{1}{\tau}\left[-i(t)+gv(t)\right] \tag{1a}$$

where $\tau = gL$.

Equation (1a) is a linear first-order differential equation that has the same form as the rate process expressed in the Hodgkin–Huxley (HH) formulation (1952) for the n parameter associated with potassium conductance. That is, under the condition of constant voltage, the HH equation for the n parameter is

$$\frac{dn(t)}{dt} = \left[1-n(t)\right]\alpha_n - n(t)\beta_n \tag{2}$$

or

$$\frac{dn(t)}{dt} = \frac{1}{\tau_n}\left[-n(t)+n_\infty\right] \tag{2a}$$

where

$$n_\infty = \frac{\alpha_n}{\alpha_n + \beta_n} \quad \text{and} \quad \tau_n = \frac{1}{\alpha_n + \beta_n}$$

The similarity of the form of equations (1a) and (2a), for a constant voltage condition, shows that a series gL circuit produces the same kinetic description (differential equation) as the one for the linearized HH potassium conductance. Therefore, *a linear, time-invariant gL circuit can duplicate the linear, time-variant behavior of the potassium conductance at any specified constant voltage*. The voltage dependence of the conductance is then taken into account as a set of values for g and L for any given set of voltages. Alternatively, the linear, time-variant behavior of potassium conductance, at any specified constant voltage, produces responses that are indistinguishable from a gL circuit. This, then, is the basis of the inductive reactance first described by Cole and Baker.

In order to obtain an expression for the admittance, which is a frequency-domain function, equation (1a) subjected to a Laplace transform (see Cheng, 1959) yields the algebraic expression

$$j\omega I(j\omega) = \frac{1}{\tau}\left[-I(j\omega) + gV(j\omega)\right] \tag{3}$$

where $j = (-1)^{1/2}$ is an imaginary quantity, capital letters denote transformed variables, and ω is frequency in radians per second.

The admittance is then defined as

$$Y(j\omega) = \frac{I(j\omega)}{V(j\omega)} \tag{4}$$

For the gL circuit example, then

$$Y(j\omega) = \frac{g}{1 + j\omega\tau} \tag{4a}$$

Admittance is a linear, steady-state concept, and consequently, when it is used to study a nonlinear conduction process such as in membranes, the amplitude of the perturbing function must be small enough to elicit only a linear response. The term "small signal" usually denotes this type

H. M. Fishman, L. E. Moore, and D. Poussart

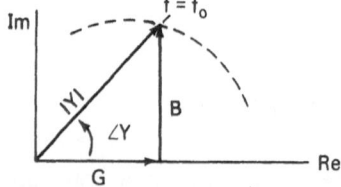

Figure 3. Admittance is expressed in the complex plane as a vector at any given frequency $f=f_0$. It can be written in rectangular form as $G+jB$ or in polar form as $|Y|\exp(j\angle Y)$.

of analysis. A "steady state" is interpreted to mean that a sufficient interval of time has elapsed after the application of changes in membrane potential so that the transient response is negligible.

The admittance may be viewed as a vector quantity in the complex plane (Figure 3) drawn from the origin to a point that corresponds to the admittance at a particular frequency $f=f_0$. The curve formed by the locus of all frequency points in the complex plane is then the admittance function. As illustrated in Figure 3, the admittance can be expressed in either of two forms. The rectangular form is

$$Y(j\omega)=G(\omega)+jB(\omega) \qquad (5)$$

where $G(\omega)$ is the real part of Y (written $\mathrm{Re}\,Y$) or horizontal component of the vector, and $B(\omega)$ is the imaginary part of $Y(\mathrm{Im}\,Y)$ or vertical component of the vector. Both $G(\omega)$ and $B(\omega)$ are, however, real functions of frequency. For the gL circuit example the rectangular form is

$$Y(j\omega)=\frac{g}{1+\omega^2\tau^2}[1-j\omega\tau] \qquad (5a)$$

Alternatively, the admittance is written in the equivalent polar form as

$$Y(j\omega)=|Y|\exp(j\angle Y) \qquad (6)$$

where

$$|Y|=(G^2+B^2)^{1/2} \quad \text{and} \quad \angle Y=\tan^{-1}(B/G)$$

For the gL circuit example the polar form is

$$Y(j\omega)=\frac{g}{(1+\omega^2\tau^2)^{1/2}}\exp(-j\tan^{-1}\omega\tau) \qquad (6a)$$

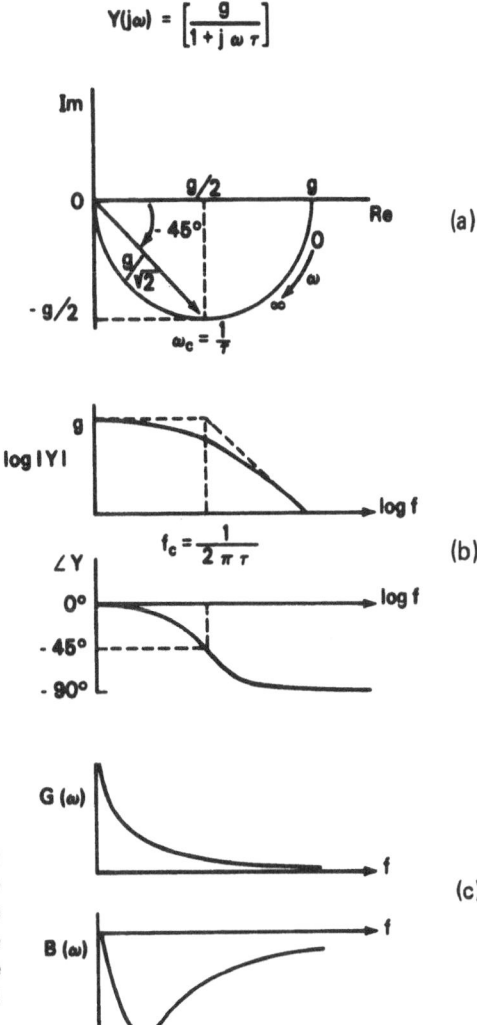

$$Y(j\omega) = \left[\frac{g}{1+j\omega\tau}\right]$$

Figure 4. Three ways of plotting complex admittance data for the series gL circuit example in Figure 2c. (a) Complex plane (Cole–Cole plot). (b) Bode plots, magnitude and phase vs. frequency. (c) Real and imaginary parts vs. frequency.

There are three ways of presenting admittance data, illustrated in Figure 4 for the gL circuit example. In the upper plot, the admittance locus is plotted in the complex plane. This type of presentation was analyzed and used by Cole (1928) and has come to be known as a Cole–Cole plot after the work of K. S. Cole and R. H. Cole (1941). A second presentation (middle portion of Figure 4) is in terms of two plots as a function of $\log f$; namely, (1) logarithm of the magnitude of admittance, $\log|Y|$, and (2) the phase angle of the admittance, $\angle Y$. These are known as Bode (1945) plots. Finally, in the lower portion of Figure 4, the real and imaginary parts of Y, i.e., $G(f)$ and $B(f)$, are plotted as separate functions of frequency. Although all of these presentations contain the same information, each has advantages as well as disadvantages that depend on the specific way in which data are collected as well as the way they are going to be used.

The Cole–Cole plot is the most compact presentation because it contains all of the information in a single plot. It also facilitates manipulations such as the subtraction of one admittance from another since this must always be done through vector operations involving the real and imaginary part of each admittance. However, it has a significant disadvantage during experimental manipulation when two or more admittance functions are measured or compared under different measurement conditions for which the Re Y and Im Y change substantially. The scales of the Re and Im axes must be chosen in advance or by trial and error in order to accommodate the largest values of Re Y and Im Y. Consequently, there is less efficient use of experimental time and/or a loss of detail for those Y functions that have a small Re or Im part relative to the scales chosen.

The Bode plot eliminates this difficulty through a log scale for the magnitude and the ability to express any phase angle within the fixed limits of $\pm 180°$. In addition, the $|Y|$ function of frequency is useful in spontaneous current noise studies since it, together with the background voltage noise of the measurement system, allows correction for background current noise. Furthermore, power density spectra of fluctuations are conveniently presented as log magnitude-squared versus $\log f$, which permits a direct comparison with a $\log|Y|$ versus $\log f$ plot. Since the above two points are important in this paper, the Bode presentation is the one that is used.

In the third presentation plots of $G(\omega)$ and $B(\omega)$ have the aforementioned scaling problem encountered in Cole–Cole plots. Under certain circumstances $G(\omega)$ and $B(\omega)/\omega$ have been interpreted literally as reflecting "membrane conductance and capacitance," respectively, and exclusively. Takashima and Yantorno (1977) have made use of this interpretation previously and Taylor *et al.* do so in this volume (Chapter 5). However, as discussed earlier, time- and voltage-dependent conductances in membranes display inductive-like as well as capacitive-like behavior and contribute to both $G(\omega)$ and $B(\omega)/\omega$. Consequently, whenever $B(\omega)/\omega$ data are used to make inferences about dielectric phenomena in membranes, contributions from ion conductance behavior must be expected unless all conductances have been eliminated.

PREPARATION AND LOW-NOISE VOLTAGE CLAMP TECHNIQUE

The preparation was the isolated giant axon of squids (*Loligo pealei*), which were received live at the Marine Biological Laboratory in Woods Hole. Most of the methods used have been described previously (Fishman *et al.*, 1977b). A "piggyback" type internal electrode was inserted axially from a cut made in one end of the axon. The electrode was electrically insulated at the cut end. Air gaps, 6 mm in length, on both sides of a central measuring region provided attenuation of the current from the axon ends, which were deactivated by heating with a nichrome wire. The central region was 4 mm in length and contained flowing artificial seawater (ASW). An air-gap method was used instead of a guarded-current scheme because of the unfavorable noise conditions presented by the latter (see Fishman, 1981). The central region contained a Ag–AgCl potential reference electrode and a Pt–Pt ground sheet in addition to a thermocouple probe for temperature measurements. The only significant modification (for low-noise operation) in the voltage clamp circuit over those used previously was in the primary (membrane potential measuring) stage. A differential amplifier was constructed from a pair of 8 paralleled field-effect-transistor input, operational amplifiers (selected National LF356). The overall output voltage of this stage was the amplified ($100\times$) difference between the sum of $1/8$ of each of the individual operational amplifier outputs. The overall intrinsic voltage noise from this stage was thus reduced by the factor 0.35 ($1/8^{1/2}$) times that attainable by use of a single LF356 as a differential input stage. The first stage gain ($100\times$) renders the noise from all subsequent stages insignificant relative to the first stage noise. The command potential to the control amplifier was then 100 times the potential that was actually applied to the membrane. This scheme reduces any extraneous noise associated with the command signals by $100\times$.

FOURIER SYNTHESIZED PSEUDORANDOM SIGNAL (FSPS)

In our development of a system for rapid measurement of admittance, the perturbation signal used in the measurement has evolved from Gaussian "white" noise (Fishman, 1975, Fishman *et al.*, 1977b) to a pseudorandom binary signal (PRBS) (Poussart *et al.*, 1977; Fishman *et al.*, 1979) to the present one which is a Fourier-synthesized, pseudorandom signal (FSPS) (Nakamura *et al.*, 1977). The equation for this signal is

$$f(t) = \sum_{k=1}^{400} A_k \sin(2\pi km/1024 + \phi_k)$$

$$m\Delta T \leqslant t \leqslant (m+1)\Delta T \tag{7}$$

where $\Delta T = T = T/1024$, T is the signal period, and $m = 0, 1, 2, \ldots, 1023$.

Four examples of an FSPS are shown in Figure 5. All four were generated in the same way, and we focus on a description of one of them, "white" noise, for brevity. Since white noise contains uniform energy at all frequencies, the function $f(t)$ was generated by a digital computer with equal Fourier amplitudes ($A_k = 1$) and a random set of values for the phases ϕ_k. The number of sinusoidal frequency components was set at 400 because this is the number of frequency points that are obtained from the fast Fourier transform (FFT) implemented in our system. For any given set of ϕ_k values, the time function $f(t)$ was plotted as 1024 equally spaced time points. One particular function (upper plot in Figure 5a) out of 30 computed was chosen on the basis of minimum peak-to-peak amplitude excursion. This selection criterion was used in order to obtain a test signal that when applied to the membrane would produce minimal excursion (dynamic range) of the controlled variable. This property is important in minimizing computational errors because of the finite dynamic range of the digital signal processing that occurs in the analysis, as well as minimizing nonlinear responses from the preparation. Each of the 1024 digital values of the selected time function $f(t)$ is permanently stored in a read-only-memory, and each is retrieved sequentially ("clocked out") by application of a repetitive pulse train (the Fourier-processor sample clock, Figure 8) to the memory device at a rate such that the generated time waveform has frequency components which lie predominantly within the frequency analysis range. The resulting staircase waveform is "smoothed" by low-pass filtering to obtain an analog perturbation signal that is deterministic (has fixed and known spectral properties in the frequency range of interest) and can be applied as a repetitive and continuous command to a voltage clamp system.

An important advantage of the FSPS over the previously used pseudorandom binary sequence (PRBS) (Poussart *et al.*, 1977) is that

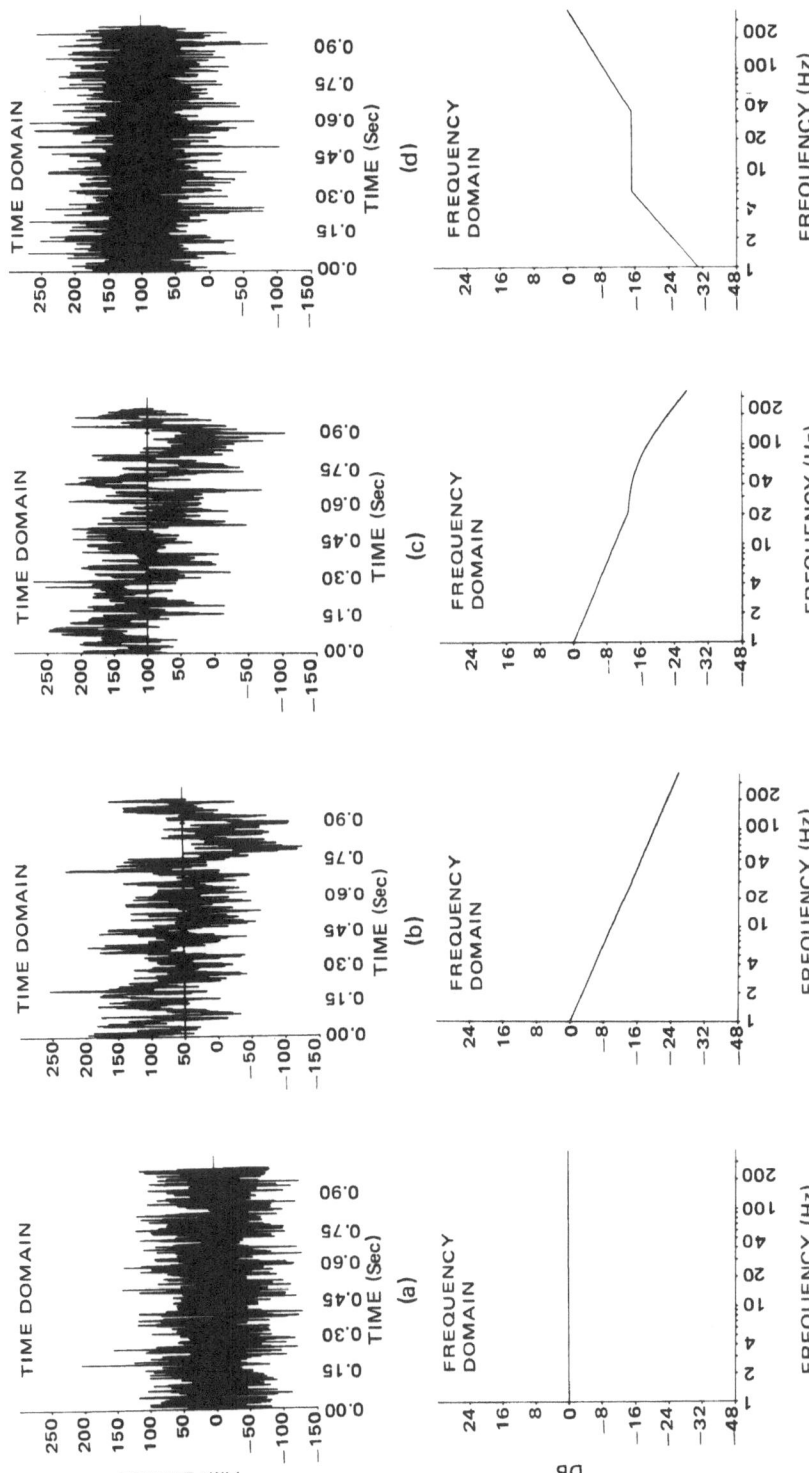

Figure 5. Four Fourier-synthesized pseudorandom signals (FSPS) that are used as small perturbations superposed on step voltage clamps to measure axon admittance. Upper: 1024 points of a digital (time) sequence. Lower: computed frequency spectrum of each upper digital sequence. See text for details.

synchronization is simplified by direct use of the Fourier-processor sample clock to run the FSPS. With the PRBS method, the period of the PRBS must be precisely matched to the duration of the time window of the FFT sampling (Poussart and Ganguly, 1977) which requires a 10 MHz master clock that limits the frequency analysis to 10 kHz or less. With the FSPS the upper frequency limit of the measurement is dependent only on the highest frequency of operation of the analog-to-digital converter. The particular FFT processor we used (Rockland model 512S) provides a 400-point spectrum in 12 analysis bands from 20 Hz to 100 kHz in a 1–2–5 sequence. The frequency resolution in each band is then the highest frequency/400 and the required length of data necessary for one spectral analysis is the reciprocal of the frequency resolution. Thus the highest frequency of operation is 100 kHz with a resolution of 250 Hz and a data length of 4 msec.

The other advantages of the PRBS method (Poussart and Ganguly, 1977), namely, fixed spectral properties independent of time base, and sequential application of stimulus and response through the same anti-alias filter with automatic removal of the filter characteristics, also apply to the FSPS.

Since the magnitude of impedance or admittance, over several decades in frequency, can easily be beyond the dynamic range of the digital processing instrumentation, the other three FSPS functions shown in Figure 5b–5d (upper plots) were used at various times as the test perturbation in order to "whiten" the response and reduce errors due to finite dynamic range. The nonuniform spectral distribution (Figures 5b–5d, lower plots) of these Fourier-synthesized stimulus signals was automatically canceled in the transfer function computation since their properties are invariant from stimulus to response.

A METHOD OF COHERENCE ELIMINATION

A significant problem arises in the analysis of squid axon current fluctuations and admittance during the time course of long-duration, step clamp currents. In the upper trace of Figure 6, the potassium current step response rises initially to a maximum value and then declines slowly during a 625-msec pulse. This behavior is the well-known current "droop." Spectral analysis of the current fluctuations during the droop includes

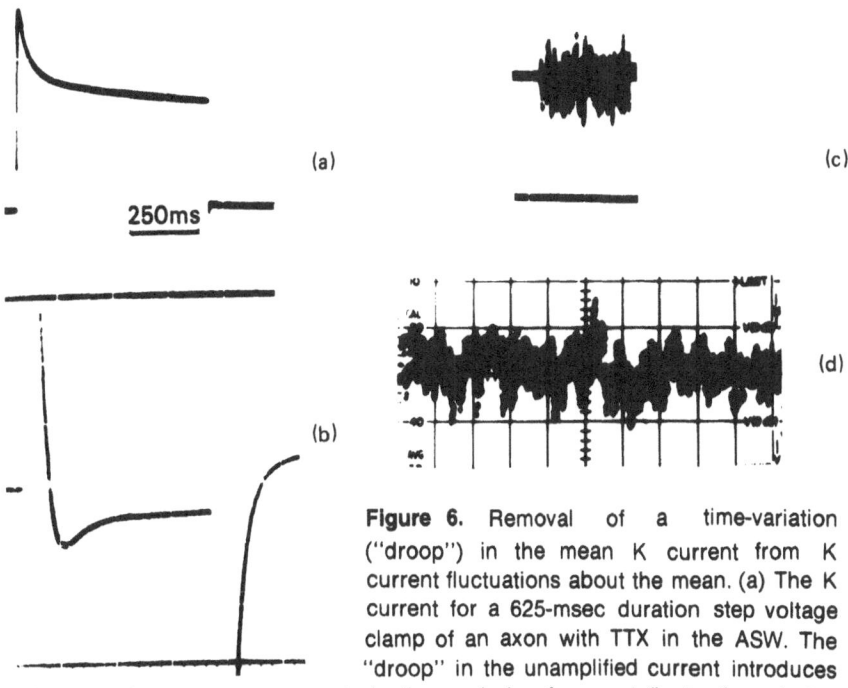

Figure 6. Removal of a time-variation ("droop") in the mean K current from K current fluctuations about the mean. (a) The K current for a 625-msec duration step voltage clamp of an axon with TTX in the ASW. The "droop" in the unamplified current introduces extraneous frequency components in the analysis of current fluctuations during the time indicated by the 250-msec bar. After high-pass filtering, shown in (b), the pedestal about which the fluctuations occur is reduced, allowing amplification without saturation. After coherence elimination (c), the current droop is removed from the fluctuations during the 250-msec measurement interval (see text). The noise sample record after analog-to-digital conversion in the Fourier processor instrument is shown in (d).

contaminating spectral components from the time variation of the current as well as the desired spectral components associated with the potassium conductance. In order to remove the contaminating spectral components from the analysis, a procedure for coherence elimination was implemented for both the admittance and noise analysis. First the pedestal about which the current fluctuations occur or on which the response to an applied FSPS occurs is removed with a simple, passive, high-pass filter prior to amplification in order to obtain the maximum dynamic range in the analysis of fluctuations or the FSPS response. In the case of an admittance measurement, Figure 7, the polarity of a small amplitude FSPS, which is superposed on the step clamp command, is alternatively changed ("toggled") with each successively applied step

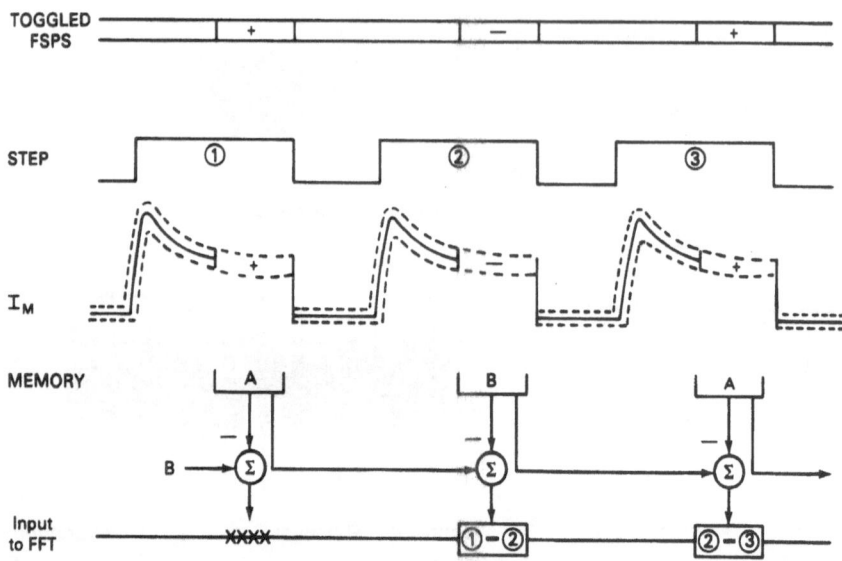

Figure 7. A method for eliminating the coherent (droop) portion of the current response to repetitive step voltage clamps from admittance or noise data, as described in Figure 6. In an admittance measurement the FSPS perturbation is toggled (waveform polarity changed) on each successively applied clamp pulse. Thus a subtraction of the digitized current response to the FSPS in memory B (for step 2) from the digitized response in memory A for (step 1) results in a reinforcement of the response while the current "droop," which is coherent from one pulse response to the next, is eliminated. In noise analysis the FSPS is not applied and the spontaneous current fluctuations, which are not coherent from one pulse to the next also reinforce while the time variation in the current (droop) is eliminated. For an admittance analysis a single pair of pulses is sufficient for a measurement, whereas in noise analysis a power spectrum is computed from the result of each subtraction of adjacent pulse data (1−2, 2−3, etc.) until a desired number of power spectrum averages is obtained. Thus slow changes in the current droop waveform are tracked and eliminated since each subtraction is performed on data which are closest in time.

clamp. The high-pass-filtered, amplified current, I_M, in response to the first step clamp is digitized at the desired time interval during the first step and placed in memory A. The polarity of the FSPS is changed prior to application of the second step clamp and the procedure repeated with the digital values of the response placed in another memory, B. Immediately after memory B has been loaded, a digital subtraction is performed between the corresponding digital values in memory A and those of memory B, and the resulting digital sequence is then converted to analog form and applied to the Fourier processor for analysis. This procedure is repeated until the measurement is completed.

Usually the admittance is obtained from the response to a single pair of FSPS-superposed, step-clamp pulses, and there is no need for further data unless the average of several admittance determinations is desired. The toggled (alternating polarity) FSPS makes the response, which carries the admittance during any successive pair of step clamps, non-coherent. Thus if the droop remains coherent (the same) during the acquisition of data for each step, it will be eliminated from the record obtained by subtraction of any successive pair of digital data sequences, while the responses to the FSPS will add to double the admittance magnitude, which is easily corrected. In the case of noise measurements, the perturbation signal is not applied and the digital values in memory reflect both the fluctuations, which are incoherent with respect to each applied pulse, and the slow droop of current, which is coherent (the same) during each step clamp. The pairwise subtraction effectively eliminates the droop from the digital data sequence upon which an FFT is computed. Thus by adding one pulse to the number of step-clamp pulses used to acquire the noise data, the spectral contamination from the droop is prevented. An advantage of this method over one in which the mean of all current responses is subtracted from each individual current response (Sigworth, 1977) is that the subtraction is done pairwise on digitized records that are close in time. Thus the effects of slow changes in the droop are tracked in time and produce less spectral distortion.

THE MEASUREMENT SYSTEM

The entire measurement system is illustrated in Figure 8. In the admittance mode, a request for a measurement is made by the momentary closure of a switch in a sequencer. The sequencer (a programmed microprocessor) continuously monitors the FSPS and computes the period of the FSPS. A premeasurement interval, which is the time after the onset of a step prior to the admittance measurement, is set by means of switches. The sequencer then computes the time at which the step should be applied (which takes into account the FSPS period) to the clamp system in order to achieve the desired delay after the onset of the step, and issues a trigger to a pulse generator (step command) at the appropriate time. Just prior to the measurement interval, the sequencer also triggers the coherence eliminator to begin its analog-to-digital (A/D) conversion and storage of data, and it closes a switch at the input to the

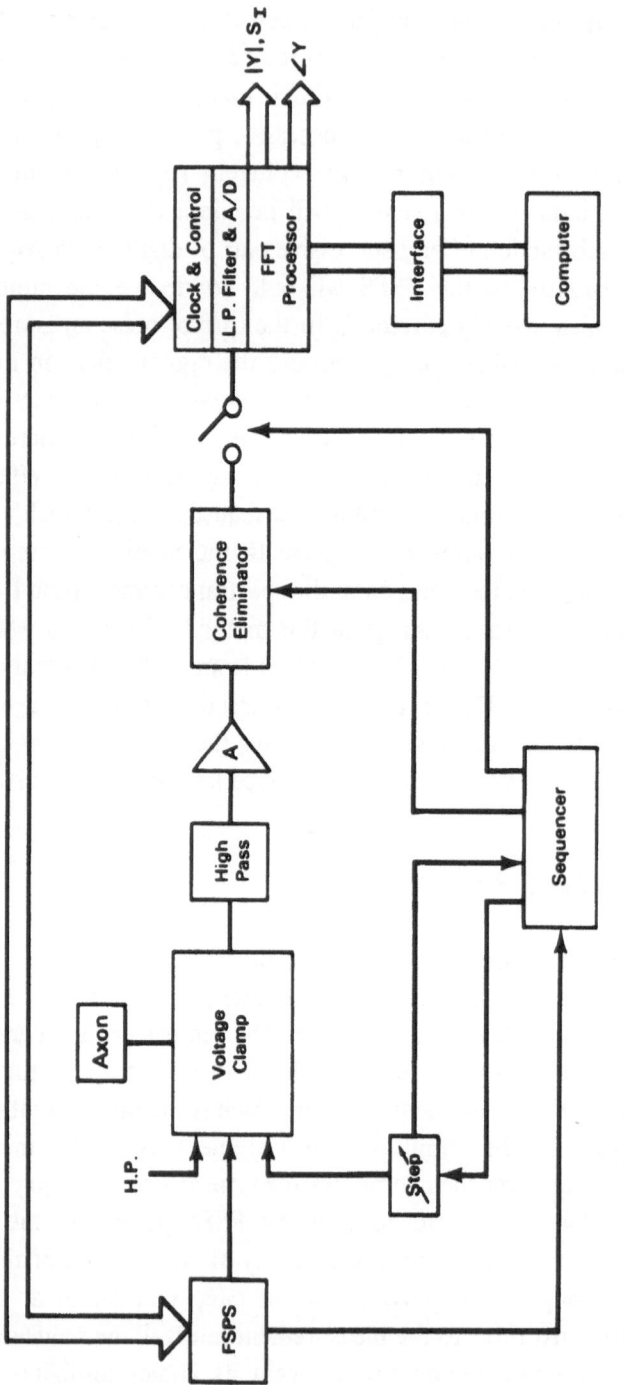

Figure 8. A schematic diagram of the entire measurement system (see text for details).

Fourier processor to allow the data to flow into the processor. The sequencer also monitors the pulse generator output to determine when the step command has terminated so that the number of pulse measurements can be varied from one (single-shot mode) to any desired number (repetitive mode).

The Fourier processor is a commercially available instrument which contains antialias filters (120-dB/octave roll-off) for each frequency analysis range. Analog signals are applied directly to the input of the Fourier processor since it contains an A/D converter. The sample clock in the processor runs at 2.56× the highest frequency on any analysis range; its output runs the FSPS continuously. The sample-clock pulses are then returned from the FSPS slightly delayed (10 μsec) in time and used to run the A/D of the Fourier processor. Use of the processor sample clock as the master timer assures precise synchronization between the FSPS, sequencer operations, and data conversion by the Fourier processor, all of which are set by a single control—the analysis range switch. Measurements of spontaneous current fluctuations are carried out in the same manner without the application of the FSPS to the clamp system.

The FFT processor is interfaced to a Digital Equipment Corporation LSI-11 minicomputer and floppy disk system. The computer is used only for computation of admittance and power spectra and data storage. A typical procedure for an admittance measurement is as follows: The voltage clamp system is set to clamp a 5-kΩ resistor. One of the four FSPS functions in Figure 5 is selected, depending on the dynamic range and shape of the expected response from the axon. An analysis range is chosen as is the high-pass filter (Figure 8) cutoff frequency (1, 10, or 100 Hz), amplifier gain A, and a low-pass filter cutoff setting for the FSPS. A measurement is initiated on the resistor. The computer obtains the real and imaginary parts of the response of the clamp and signal processing system on the resistor and computes both $|Y_r|$ and $\angle Y_r$ at 400 frequency points. These functions are then stored in the computer memory as reference functions. An axon is then voltage clamped and another measurement initiated. The computer again obtains the real and imaginary parts of the axon current response and computes $|Y_{ax}|$ and $\angle Y_{ax}$ of the axon response. Then the ratio of the magnitudes $|Y_{ax}|/|Y_r|$ and the difference in phase angles $\angle Y_{ax} - \angle Y_r$ are computed at each of the 400 frequency points and both functions are displayed on a CRT as $\log|Y|$

versus $\log f$ and $\angle Y$ versus $\log f$. In this way the resulting admittance functions are automatically corrected for the nonuniform magnitude and nonzero phase properties of the perturbation function as well as those of the clamp system and the signal processing (filters, amplifiers, and all other devices) leading to the FFT process. A calibration is obtained by again clamping the 5-kΩ resistor with the same gain settings used during the axon measurement. Since this yields $|Y|$=const and $\angle Y$=0 at all frequencies, the level of $|Y|$ obtained on the display is simply the reciprocal of 5-kΩ and the $\angle Y$ function indicates the zero degrees axis.

For a noise measurement, a magnitude transfer function of the clamp system (on a 5-kΩ resistor) and the signal processing path is obtained by computing the ratio of the magnitude function of the response of the entire measurement system (for an applied FSPS) to the magnitude function of the FSPS itself. This magnitude transfer function is stored in the computer memory, and is used to correct the average of a number of power spectra of axon current fluctuations for the passband characteristic of the measurement system. During the data acquisition interval of each clamp pulse, the Fourier processor transfers the real and imaginary parts to the computer which calculates both the magnitude spectrum and the average of all spectra that have been accumulated up to the present step clamp. The passband-corrected average of all accumulated spectra is then displayed on a CRT as a power spectrum. Calibration of power spectra is done in absolute terms by calculating the current noise power density, referred to the preparation, taking into account all signal gains and the full scale sensitivity of the input section of the Fourier processor.

THE AMPLITUDE RANGE FOR A LINEAR RESPONSE

As discussed earlier, admittance is a linear concept. The applicability of this concept must be justified by demonstrating that an essentially linear response is elicited for the perturbation amplitudes used. In recent work (Moore *et al.*, 1980), we have explored the amplitude range of the linear response in the squid axon. Figure 9 shows the results of a harmonic analysis. A sinusoid was applied as a command to a voltage-clamped axon and the current response to the sinusoid was spectrum analyzed. The harmonics generated by the potassium conduction system,

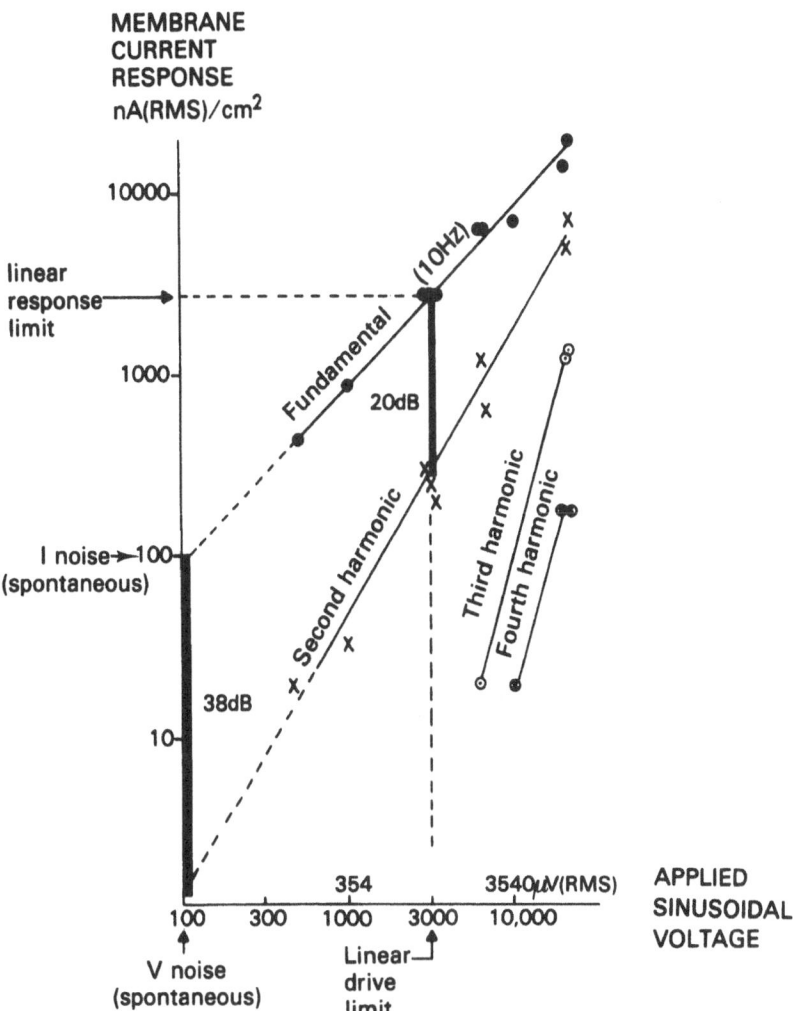

Figure 9. Harmonic analysis of a squid axon membrane current in response to a sinusoidal (10-Hz) voltage clamp of various amplitudes to explore the range of a linear response. Applied sinusoidal voltage is in μV (peak to peak), V is at clamp conditions. For amplitudes in the range from spontaneous noise up to 3 mV (peak to peak) the total harmonic content (second harmonic) is ⩽10% (−20 dB) of the fundamental in the response. From Moore et al. (1980), by permission of the *Journal of Membrane Biology*.

as the peak-to-peak (p–p) amplitude of the sinusoid was increased from 400 μV to 10 mV, gives a sensitive measure of the development and extent of nonlinearities. From Figure 9, the total harmonic content for amplitudes up to 3 mV (p–p) is in the second harmonic and represents a 10% (-20 dB) distortion relative to the amplitude of the fundamental in the response. We have arbitrarily chosen this amplitude as an upper limit for applied perturbations in order to obtain a reasonably linear response. In most experiments, however, perturbation amplitudes of 0.5 to 1 mV (p–p), well below this upper limit, yielded good signal-to-noise performance and were used for the admittance measurements.

POTASSIUM CONDUCTION IN THE CALCULATED ADMITTANCE OF SQUID AXON

There has not been extensive use of membrane admittance to describe ion conduction. It is therefore appropriate to provide a discussion of some of the essential features that are landmarks in the interpretation of data. For this purpose, we use calculations of membrane admittance from the linearized Hodgkin–Huxley (lHH) equations (see Fishman *et al.*, 1977a). The lHH admittance for the potassium conduction process is given by two terms as

$$Y_K = \bar{g}_K n_\infty^4 + \frac{g_n}{1 + j\omega\tau_n} \tag{8}$$

where $\bar{g}_K$ is the maximum potassium conductance and

$$g_n = 4\bar{g}_K n_\infty^3 \tau_n (V - V_K)\left[\frac{d\alpha_n}{dV} - n_\infty\left(\frac{d\alpha_n}{dV} - \frac{d\beta_n}{dV}\right)\right] \tag{9}$$

The first term in equation (8), $\bar{g}_K n^4$ is the chord ("infinite frequency") conductance, and the second term, $g_n(1 + j\omega\tau_n)^{-1}$, is the frequency domain manifestation of the time-variant property of potassium conductance. In Figure 10 both of these terms are realized in a circuit with and without an ideal capacitance, C, which accounts for the response of the membrane dielectric. Calculations over the frequency range of 1–3000 Hz of the magnitude and phase of potassium admittance, Y_K, as well as with the added capacitance, $Y_K + Y_C$, are plotted at various potentials

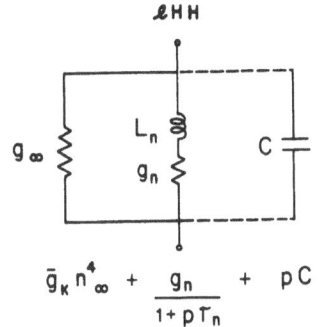

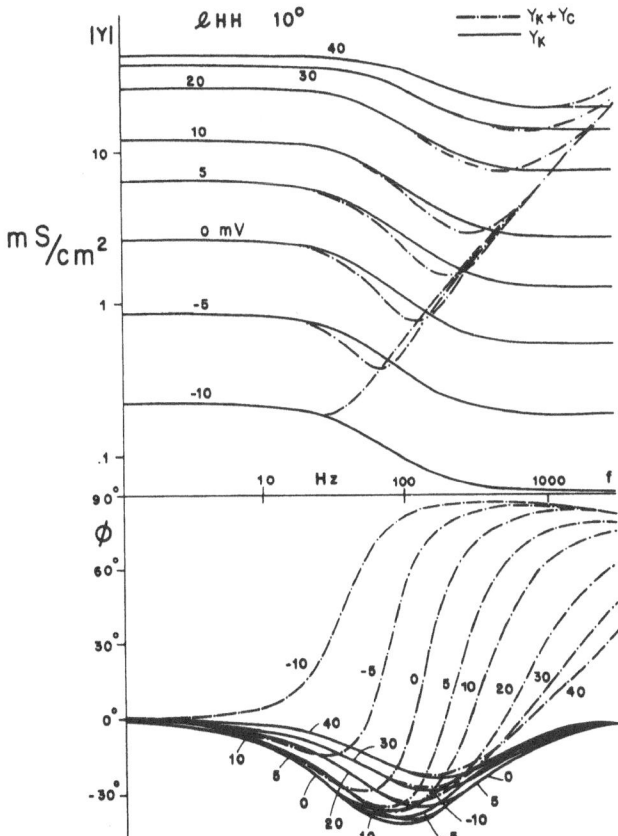

Figure 10. Circuit realization of the linearized HH K conductance and calculations of the admittance magnitude $|Y|$ and phase angle ϕ as functions of frequency with and without an assumed ideal capacitance C. A series resistance of $7\ \Omega\,cm^2$ is also included in the calculations. After Fishman *et al.* (1979), by permission of the *Journal of Membrane Biology*.

away from membrane rest-potential in Figure 10. The $|Y_K + Y_C|$ (dashed curves) show a low-frequency asymptotic value, at any given potential, that is essentially $\bar{g}_K n_\infty^4 + g_n$. At higher frequencies (50–1000 Hz) a local minimum occurs in $|Y_K + Y_C|$ that is known as an antiresonance, which can be understood as an interaction between the admittance behavior of the gL branch and the C branch. The phase function ($\angle Y_K + \angle Y_C$) first goes negative with increasing frequency and then makes a sharp transition very near the antiresonant frequency to a positive phase. It is clear that the antiresonant frequency shifts to higher frequencies as membrane potential increases. This is due to changes in the values for g_n and L_n ($= \tau_n/g_n$) with potential. The quality or sharpness of the resonance also changes with potential. At hyperpolarized potentials (e.g., -10 mV) approaching the potassium equilibrium potential (V_K), the antiresonance is diminished and it disappears (not shown in Figure 10) entirely at $V_M = V_K$, where $L_n \to \infty$ because $g_n \to 0$ [see equation (9)]. For depolarized potentials, the antiresonance becomes very broad, reflecting the increasing value of $\bar{g}_K n_\infty^4$, which "damps" the antiresonance.

COMPARISON OF POTASSIUM CONDUCTION KINETICS FROM ADMITTANCE AND NOISE DATA

It is of interest to compare the behavior of $|Y_K|$ with potential, Figure 10, (solid curves) with the total admittance magnitude, $|Y|$. The behavior of $|Y_K|$ is of interest because its functional form should be the same as the power spectrum of potassium current fluctuations, at the same constant voltage, if the fluctuations reflect a linear kinetic process and the noise generating process is "white" over the frequency range considered. Thus a comparison of the form of power spectra of potassium current noise with $|Y_K + Y_C|$, at the same membrane potential and on the same axon, provides the information from which it may be determined whether or not potassium conduction is a linear kinetic process at the level of spontaneous fluctuations. As noted above, the behavior of $|Y_K|$ below frequencies that approach the antiresonance in $|Y_K + Y_C|$ is almost identical to that of $|Y_K + Y_C|$ since $|Y_K| \gg |Y_C|$ at these frequencies. In Figure 10 $|Y_K|$ declines from its low-frequency asymptotic value as a Lorentzian function, $(1 + f/f_c)^{-1}$, and approaches another asymptotic value at high frequencies that is determined by the value $\bar{g}_K n_\infty^4$ at any given potential.

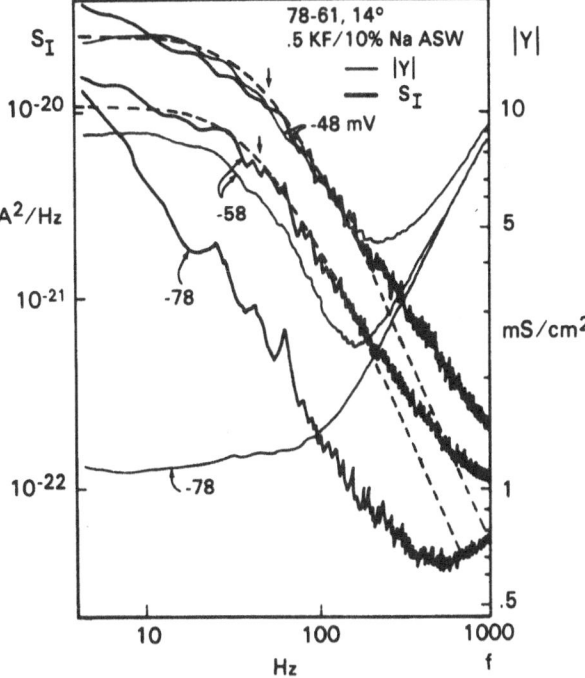

Figure 11. Comparison of the power spectrum, S_I, of K current fluctuations at three membrane potentials with the magnitude of admittance, $|Y|$, at the same potentials and in the same axon. Dashed curves are Lorentzian fits of S_I at -58 and -48 mV, with arrows indicating corner frequency. Axon area $\simeq 0.076$ cm^2.

Figure 11 contains a comparison of these two types of data from a squid axon. Sodium conduction was suppressed by 90% replacement of sodium in the external artificial seawater by Tris (TTX gave similar results). At an absolute potential of -78 mV, $|Y|$ shows only a slight indication of an inductive component in the admittance (sharpness in the corner region) and the current–noise power spectrum, S_I, shows a general decline that has the form $1/f$. The low-frequency asymptote of the total admittance magnitude, $|Y|$, at -78 mV in Figure 11, reflects essentially leakage conductance, which is about 1 mS/cm^2. Since the low-frequency admittance at -58 and -48 mV in Figure 11 is much greater, leakage is assumed to be constant with frequency and have negligible effect on the form of $|Y|$ at depolarized potentials in these and subsequent data. At rest potential (-58 mV) and for a depolarization to -48 mV, $|Y|$ shows an antiresonance, which shifts to higher frequency for the depolarization as in the dashed curves of $|Y_K + Y_C|$ in Figure 10 for the corresponding potentials of 0 and 10 mV. The data for S_I at rest and -48 mV both have essentially Lorentzian forms that match qualitatively the form of $|Y|$ at these potentials, at frequencies below the antiresonance. Both Lorentzians were replaced by lower-intensity non-Lorentzian spectra when the internal perfusate was changed from $0.5M$ to $0.05M$ KF. Therefore, they reflect potassium current noise. Similarly, the antiresonance in $|Y|$ at rest and at -48 mV is eliminated after potassium conduction is suppressed (in addition to suppression of Na conduction by TTX) (see Fishman *et al.*; 1977a), which indicates that $|Y|$ in Figure 11 reflects $|Y_K + Y_C| \simeq |Y_K|$ at low frequencies and with TTX applied.

A qualitative comparison of S_I and $|Y|$ at three different temperatures (Figure 12) also shows a good correspondence at rest potential between the shift in the "corner" region in S_I and the shift in the low-frequency (at frequencies below the antiresonance) "corner" region in $|Y|$. In Figure 12 the "corner" region of S_I shifts from about 20 Hz at $6.5°$ to about 90 Hz at $18°$ as does the "corner" in $|Y|$ at the same temperatures. The excellent agreement between the form of K current fluctuations and $|Y|$ for potential and temperature changes suggest that the kinetics derived from the two measurements, one of which is linear, are in agreement as indicated earlier (Fishman, 1973). This result for potential and temperature changes can only be obtained if the process that produces the fluctuations is a linear kinetic process for perturbation amplitudes corresponding to those of spontaneous fluctuations. Further-

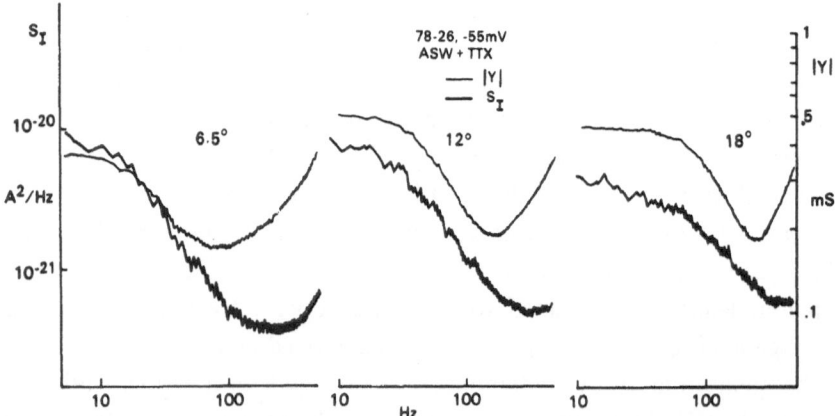

Figure 12. Comparison of S_I and $|Y|$ at rest potential in the same axon at three temperatures. Axon area $\simeq 0.06$ cm^2.

more, it is pertinent to mention that in Figure 9 an extrapolation of the harmonic response to a sinusoidal voltage clamp down to the amplitude of spontaneous noise yields a total harmonic distortion of -38 dB, about 1%. This low, and hence negligible, nonlinearity in the extrapolated response at the amplitude level of spontaneous fluctuations, is consistent with the above interpretation that admittance and noise data give the same kinetic properties. However, before this conclusion can be firmly established, it is necessary to make the comparison between admittance and noise data over a wider potential range and to unfold Y_C from Y in order to compare the spectral form quantitatively over a wider frequency range. Both of these endeavors are under way and will be the subjects of future work.

THE ADMITTANCE AND NOISE OF POTASSIUM CONDUCTION DURING SHORT- VERSUS LONG-DURATION STEP CLAMPS

In recent work by van den Berg, Siebenga, and DeBruin (1977), power spectra of potassium current fluctuations during short-duration step changes on a single node of Ranvier from frogs were completely different from the spectra of fluctuations analyzed during long polarizations. The long-term fluctuations are thought to reflect an inactivated potassium conductance in the node. In the only two noise studies in squid axon (Conti *et al.*, 1975; Fishman *et al.*, 1975), long-duration polarizations were used exclusively. It is therefore of interest to determine if long-term potassium current fluctuations in the squid axon are affected by an inactivation process.

In the first set of experiments, the admittance was measured at two different time intervals following a step voltage clamp. The analysis range was 20 kHz (resolution 50 Hz), which allowed us to keep the record length for admittance analysis to 20 msec. Consequently the measurements were made (Figure 13) at intervals $M1$ (20–40 msec) and $M2$ (1000–1020 msec) after a step change of 10, 20, and 30 mV in axons bathed in ASW plus $10^{-6}M$ TTX. The admittances in Figure 13 show the characteristic antiresonance calculated for the linearized K conductance in Figure 10 and $|Y|$ data in Figure 11. The high-frequency behavior (>5 kHz) is dominated by the capacitance and also contains some distortion, as indicated by a significant leveling off of $|Y|$ and a

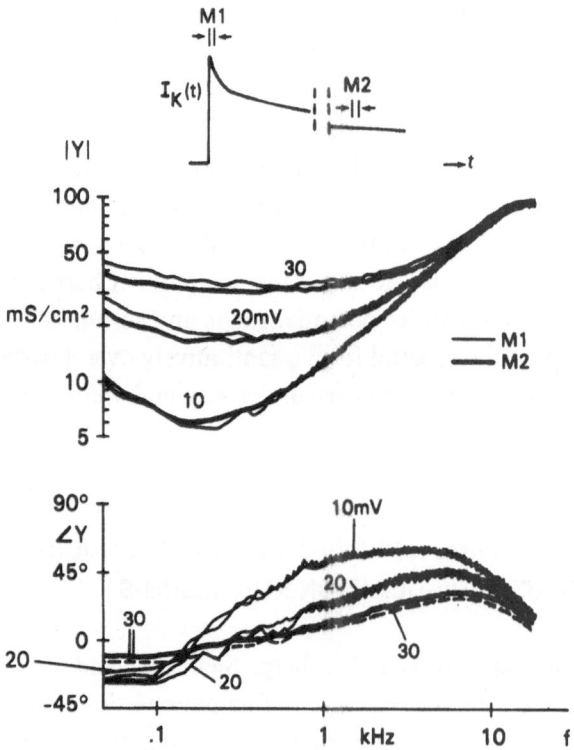

Figure 13. Comparison of the admittance of a squid axon at a short time $M1$ (20–40 msec) and a long time $M2$ (1000–1020 msec) after step clamps to three depolarized potentials from rest.

decline in $\angle Y$. The high-frequency portion of the curves is affected by the limited loop gain of the clamp system at these frequencies, which causes a distortion of the actual admittance due to an increase with frequency of the effective series resistance (the addition of the solution resistance between current and potential electrodes to the usual series resistance), which the clamp is not able to make negligible relative to the decreased membrane capacitive reactance with increasing frequency. However, the low-frequency admittance (<5 kHz), which contains the ion-conduction information, is undistorted. From the low-frequency portion of the data, the admittance is seen to be relatively the same at the two widely different measurement times, although the slight disparity appears to increase with larger depolarizations. The decrease in $|Y|$ measured at $M2$ is the maximum decrease observed because measurements at longer times (30 sec) showed no further changes from those measured at $M2$. It would appear then, from these data, that there are no major changes in the kinetics of potassium conduction in the squid axon for long-duration step clamps, during the "drooping" phase of potassium current. The decline in potassium current for long-duration step clamps probably reflects a change in the driving force for potassium ions as a consequence of the potassium ion concentration buildup in the Schwann cell space surrounding the axon.

In the second set of measurements on the same axon, the power density spectrum of potassium current fluctuations was obtained under the same measurement conditions as those in Figure 13. These data are shown in Figure 14. Again the spectra at low frequencies (<1 kHz) reflect the undistorted potassium current fluctuation properties, whereas the high-frequency portion is dominated by the background voltage noise

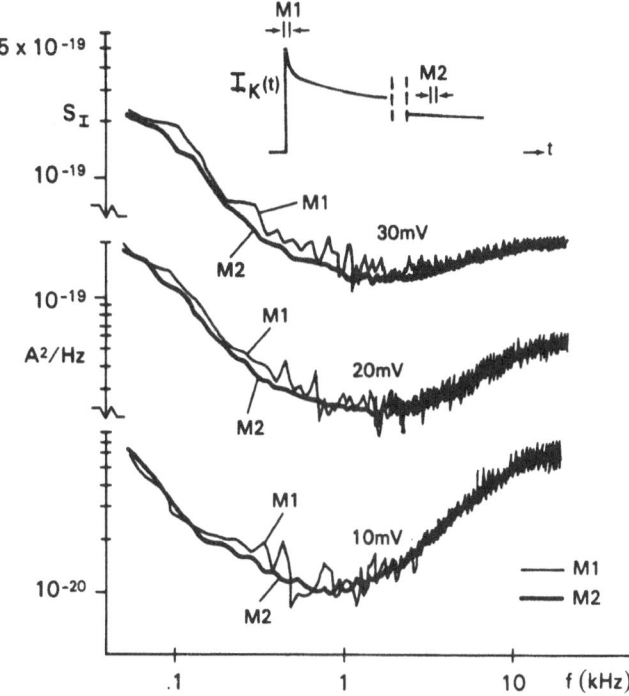

Figure 14. Comparison of power spectra of K current fluctuations in the same squid axon as Figure 13 obtained at time intervals $M1$ (20–40 msec) and $M2$ (1000–1020 msec) during step clamps to the same depolarized potentials from rest as in Figure 13. Spectra beyond 1 kHz, which have not been corrected, are dominated by fluctuations produced by background voltage noise from electrodes and membrane admittance.

of the potential electrodes and instrumentation, which is injected into the membrane admittance by the clamp system. At each of the three depolarizations from rest, there is no significant difference between the spectra of fluctuations obtained during $M1$ when compared to those obtained at $M2$. These results are thus in agreement with the measurements of admittance versus time. Together they provide strong evidence that potassium conduction kinetics are not significantly altered in the squid axon over a 30-mV potential range from rest and during long depolarizations. Consequently analyses of steady-state potassium current fluctuations in the squid axon relate directly to the characteristic time of potassium conduction kinetics. These results on the squid axon then differ from those reported for frog node of Ranvier (van den Berg *et al.*, 1977).

ACKNOWLEDGMENTS

The coherence eliminator and sequencer were expertly designed by Mr. William Law, Jr. and built by Mr. Michael Fason. Mr. David Ray implemented the computer programs for the noise signals and the on-line admittance measurements. We thank Dr. H. Richard Leuchtag for careful review of the manuscript and his helpful comments. Ms. Sarah Adams typed the manuscript versions of this article. This work was supported in part by grants NS 11764, NS 13778, and NS 13520 from the NIH and by Canadian Research Council grant A-5274.

REFERENCES

Bode, H. W. (1945). *Network Analysis and Feedback Amplifier Design* (Van Nostrand, New York), pp. 151–157.

Chandler, W. K., FitzHugh, R., and Cole, K. S. (1962). Theoretical stability properties of a space-clamped axon, *Biophys. J.* **2**, 105–127.

Cheng, D. K. (1959). *Analysis of Linear Systems* (Addison-Wesley, Reading, Massachusetts), pp. 155–180.

Cole, K. S. (1928). Electric impedance of suspensions of spheres, *J. Gen. Physiol.* **12**, 29–36.

Cole, K. S. (1947). Electrical conductance of the cell membrane, in *Four Lectures on Biophysics by Kenneth S. Cole at Instituto de Biofisica* (Universidade de Brasil, Rio de Janeiro).

Cole, K. S. (1949). Dynamic electrical characteristics of the squid axon membrane, *Arch. Sci. Physiol.* **3**, 253–258.

Cole, K. S. (1972). *Membranes, Ions and Impulses. A Chapter of Classical Biophysics* (University of California Press, Berkeley).

Cole, K. S., and Baker, R. F. (1941). Longitudinal impedance of the squid giant axon, *J. Gen. Physiol.* **24**, 771–778.

Cole, K. S., and Cole, R. H. (1941). Dispersion and absorption in dielectrics. I. Alternating current characteristics, *J. Chem. Phys.* **9**, 341–351.

Conti, F., DeFelice, L. J., and Wanke, E. (1975). Potassium and sodium ion current noise in the membrane of the squid giant axon, *J. Physiol.* **248**, 45–82.

Fishman, H. M. (1973). Relaxation spectra of potassium channel noise from squid axon membranes, *Proc. Natl. Acad. Sci. (USA)* **70**, 876–879.

Fishman, H. M. (1975). Rapid complex impedance measurements of squid axon membrane via input–output cross correlation function, in *Proceedings of the First Symposium on Testing and Identification of Nonlinear Systems*, G. D. McCann and P. Z. Marmarelis, Eds. (California Institute of Technology, Pasadena), pp. 257–274.

Fishman, H. M. (1981). Current and voltage clamp techniques, in *Techniques in Cellular Physiology*, P. F. Baker, Ed. (Elsevier/North-Holland, in press).

Fishman, H. M., Moore, L. E., and Poussart, D. J. M. (1975). Potassium-ion conduction noise in squid axon membrane, *J. Membr. Biol.* **24**, 305–328.

Fishman, H. M., Moore, L. E., and Poussart, D. J. M. (1977a). Asymmetry currents and admittance in squid axons, *Biophys. J.* **19**, 177–183.

Fishman, H. M., Poussart, D. J. M., Moore, L. E., and Siebenga, E. (1977b). K+ conduction description from the low frequency impedance and admittance of squid axon, *J. Membr. Biol.* **32**, 255–290.

Fishman, H. M., Poussart, D., and Moore, L. E. (1979). Complex admittance of Na+ conduction in squid axon, *J. Membr. Biol.* **50**, 43–63.

Hodgkin, A. L., and Huxley, A. F. (1952). A quantitative description of membrane current and its application to conduction and excitation in nerve, *J. Physiol.* **117**, 500–544.

Moore, L. E., Fishman, H. M., and Poussart, D. J. M. (1980). Small-signal analysis of K+ conduction in squid axons, *J. Membr. Biol.* **54**, 157–164.

Nakamura, H., Husimi, Y., and Wada, A. (1977). An application of Fourier synthesis to pseudorandom noise dielectric spectrometer, *Jpn. J. Appl. Phys.* **16**, 2301–2302.

Poussart, D. J. M., and Ganguly, U. S. (1977). Rapid measurement of system kinetics—an instrument for real-time transfer function analysis, *Proc. IEEE* **65**, 741–747.

Poussart, D. J. M., Moore, L. E., and Fishman, H. M. (1977). Ion movements and kinetics in squid axon. I. Complex admittance, *Ann. NY Acad. Sci.* **303**, 355–379.

Sigworth, F. J. (1977). Sodium channels in nerve apparently have two conductance states, *Nature* **270**, 265–267.

Takashima, S., and Yantorno, R. (1977). Investigation of voltage-dependent membrane capacity of squid giant axons, *Ann. N.Y. Acad. Sci.* **303**, 306–321.

van den Berg, R. J., Siebenga, E., and DeBruin, G. (1977). Potassium ion noise currents and inactivation in voltage-clamped node of Ranvier, *Nature* **265**, 177–179.

5

Squid Axon Membrane Low-Frequency Dielectric Properties

ROBERT E. TAYLOR, JULIO M. FERNÁNDEZ,
and FRANCISCO BEZANILLA

SMALL SIGNAL ELECTRICAL EQUIVALENT OF SQUID GIANT AXON MEMBRANE

On the occasion of the 65th birthday of Dr. Kenneth S. Cole, Taylor (1965) presented a discussion of measurements of the impedance of the squid axon membrane. That discussion was mainly concerned with results obtained by Taylor and Chandler (1962) from 10 to 70 kHz, the frequency range chosen because of inaccuracies of the bridge above 100 kHz and the impedance contributed by the voltage and time-dependent ionic conductances below 5 kHz.

ROBERT E. TAYLOR • Laboratory of Biophysics, NINCDS, National Institutes of Health, Bethesda, Maryland 20205.
JULIO M. FERNÁNDEZ and FRANCISCO BEZANILLA • Department of Physiology, Ahmanson Laboratory of Neurobiology of the Brain Research Institute, and Jerry Lewis Neuromuscular Research Center, University of California, Los Angeles, California.

On this occasion of Kacy's 80th birthday we present some recent results over the range of a few hundred hertz to 10 kHz which we obtained during the early part of the 1980 season at the Marine Biological Laboratory in Woods Hole.

Our discussion here is restricted to those membrane characteristics where the currents due to applied voltages are displacement currents in the (probably) nonlossy[†] dielectric of the lipid bilayer portion of the membrane and the voltage-independent displacement currents in the lossy dielectric portion (penetrating proteins?).

As shown in Figure 1 the current through the membrane in response to small amplitude voltages is considered by us to consist of several components: (1) ionic current, as represented by the linearized Hodgkin–Huxley (HH) equations (see Chandler, FitzHugh, and Cole, 1962) and an additional nonlinear, voltage-dependent leakage; (2) displacement currents in the (probably) nonlossy dielectric of the lipid bilayer portions of the membrane; (3) voltage-independent displacement currents in the lossy dielectric (penetrating proteins?); and (4) displacement currents in that part of the dielectric which is lossy and saturable and related to the ionic permeability gating mechanisms. Thus our interest here is in items (2) and (3).

ON THE MEASUREMENT OF MEMBRANE CAPACITANCE AND CONDUCTANCE

Experiments were carried out using perfused, voltage-clamped axons from the squid *Loligo pealei*. In order to eliminate the ionic currents (HH ionic, Figure 1) the permeant ions were replaced in the external solution by Tris (trishydroxy methyl amino methane) solutions and in the internal solution with tetra methyl ammonium fluoride-glutamate and sucrose and by the use of tetrodotoxin externally. The details of the experimental procedures will be reported in the near future (Fernández *et al.*, submitted for publication).

Many experiments, both in the frequency and time domain, were conducted which demonstrated that the displacement currents associated

---

[†]By definition, a dielectric is nonconducting for constant electric fields but may still dissipate heat with the application of time-varying electric fields. This is referred to as dielectric loss.

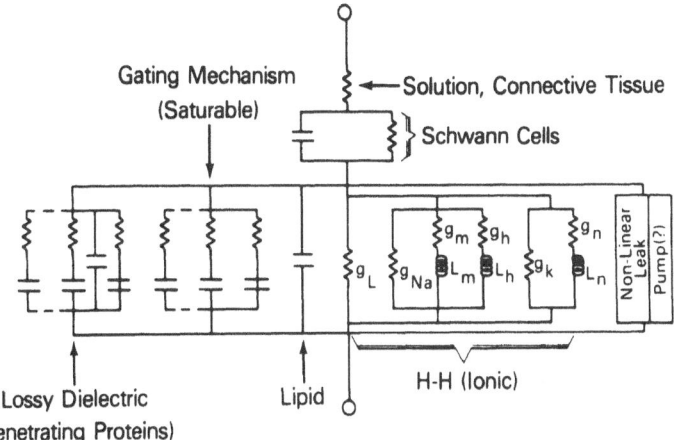

Figure 1. Equivalent electrical circuit for the squid axon membrane. It is suggested that proteins in the membrane are responsible for much of the dielectric loss other than that due to the gating current mechanisms. The latter saturate at potentials above about $+50$ mV and below about -150 mV and contribute significantly to the impedance up to about 5 kHz. The lipid bilayer portion of the membrane is thought to be nonlossy and to have a capacitance of some 0.6 $\mu F/cm^2$, referred to its area. The time- and voltage-dependent ionic currents contribute components below 5 kHz and are represented here as the linearized Hodgkin–Huxley equations. The non-linear voltage-dependent leakage becomes larger and time-dependent in old axons. The possibility of ionic pumps having an effect is indicated for completeness. The series impedance due to the Schwann cell layer is probably a pure resistance below 100 kHz and above a few hundred hertz. The possibility that the Schwann cell layer is responsible for a polarization capacity at very low frequencies cannot be ruled out at this time.

with the ionic permeability gating mechanisms (Figure 1) become very small for large hyper- or depolarizations. At $+40$ mV (internal minus external potential) or at -140 mV these currents are negligible. The results reported here were done at a holding potential of zero where the gating current contribution is small.

After an axon had been subjected to considerable experimentation it often happened that an additional current component appeared which was clearly time and voltage dependent which we consider to be an unidentified ionic leakage. When this was observed we stopped the experiment. All frequency-domain experiments were preceded and followed by time-domain measurements of gating currents to be sure that the membrane properties had not changed.

To measure admittance (or impedance) a 2-mV amplitude command sequence was applied to the voltage clamp. This consisted of the filtered output of a pseudorandom binary sequence generator (Clausen and Fernández, 1981) which could be set to cover the range up to 1.319, 3.299, or 13.196 kHz. Either the measured membrane voltage or the current through the membrane was applied, at different times, through a 120-db/octave filter and amplifier to the input of an analog-to-digital converter connected to a Nova 3 computer. These data were processed with fast Fourier transform software to provide real and imaginary parts (or magnitude and phase) of admittance or impedance. These results were then corrected for series resistance and then for parallel leakage. The series resistance correction term is still not completely objective. We measured the resistance between the voltage recording electrodes with and without the axon in place and the total series resistance ranged from 6 to 8 $\Omega\,cm^2$. Our estimate of the portion due to the Schwann cell layer, basement membrane and connective tissue (see Binstock *et al.*, 1975 for discussion and references) is similar to the value of 2.8 ± 0.6 $\Omega\,cm^2$ recently reported by Salzberg, Bezanilla, and Dávila (1980), using voltage sensitive dyes. This can be compared to the estimate of Taylor (1965) of 2.75 to 3.45 $\Omega\,cm^2$.

A few definitions should be helpful:

Admittance $= Y = G + jB =$ conductance $+ (j)$ susceptance $= I(j\omega)/V(j\omega)$
Impedance $= Z = 1/Y = R + jX =$ resistance $+ (j)$ reactance
Complex capacitance $= C_1 - jC_2 = Y/(j\omega) = B/\omega + jG/\omega = C^*$

where $j = (-1)^{1/2}$, $\omega = 2\pi f$, where f is frequency, and $I(j\omega), V(j\omega)$ are the Fourier transforms of current and voltage, respectively.

The customary way to present the complex capacitance is to plot C_2 versus C_1. This Argand diagram is often referred to as a Cole–Cole plot. C_1 is proportional to the energy stored per cycle and C_2 the energy dissipated per cycle.

MEMBRANE CAPACITANCE AND CONDUCTANCE

Measurements made at various holding potentials showed that the low-frequency capacitance reached a peak at about -60 mV (outside ground). These, and other measurements, both in the frequency and the time domain will be presented elsewhere. From these results we concluded that the voltage-dependent contribution is small at a holding

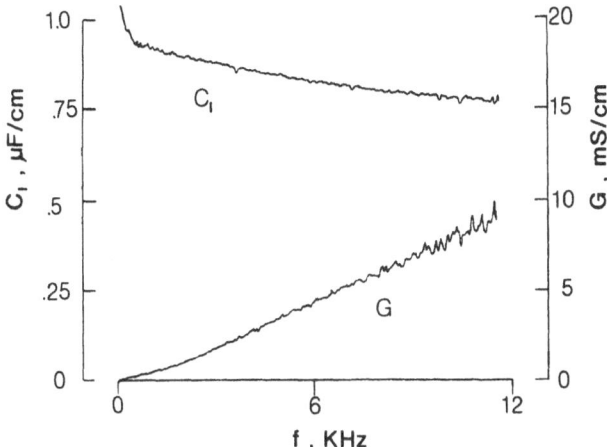

Figure 2. Experimental results for an axon held at zero millivolts potential where the contribution of gating current to the dielectric properties is small. The upper curve is the capacitance and the lower, the conductance after correction for 6.2 $\Omega\,cm^2$ series resistance, each plotted against the same linear frequency. The digitized output consisted of 512 points over a frequency range of 13.196 kHz. The data points at 60 Hz and its harmonics have been substituted by the average of the two adjacent points and those for frequencies below a few hundred hertz and above some 10 kHz are not shown because of excessive noise in these regions.

potential of 0 mV. Figure 2 shows the results of measurements at this holding potential. The capacitance C_1, and conductance are plotted as a linear function of frequency. The data have been corrected for a total series resistance (between voltage-measuring electrodes) of 6.2 $\Omega\,cm^2$ with no correction for leakage. Points below a few hundred hertz and above about 10 kHz are not shown because of lack of resolution and aliasing artifacts,[†] respectively. The leakage, i.e., the limit of the conductance for zero frequency, is not visible on this plot. However, using a 3.299-kHz bandwidth we measured a leakage of 11.67 $k\Omega\,cm^2$.

The complex capacitance for the data of Figure 2 is shown in Figure 3 along with the data obtained for a bandwidth of 3.299 kHz (corrected for a leakage of 11.66 $k\Omega\,cm^2$ and series resistance of 6.2 $\Omega\,cm^2$) and two theoretical curves. Since the values chosen for series resistance and leakage are very critical for this plot the results are presented as being

---

[†]Aliasing results in irreversible loss of data due to sampling of frequencies above the Nyquist or folding frequency, which is one-half of the sampling frequency (Bendat and Piersol, 1971).

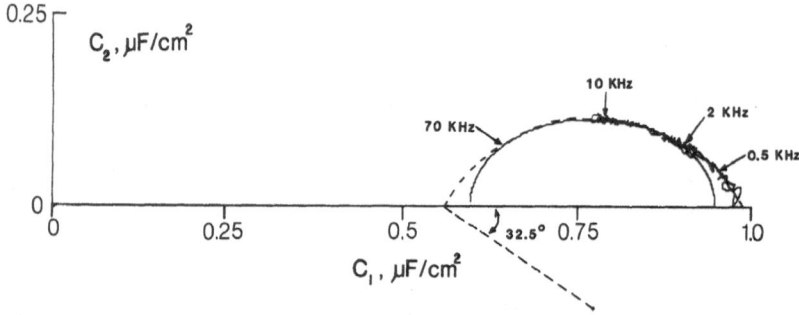

Figure 3. Imaginary part of complex capacitance (G/ω) versus (B/ω) (Cole–Cole plot). The noisy part of the locus is the superimposition of results obtained with 3.299 kHz bandwidth (corrected for 6.2 $\Omega\,cm^2$ series resistance and 11.67 k$\Omega\,cm^2$ parallel leakage) and the results shown in Figure 2 for 13.196 kHz bandwidth. Measurements were made on the same axon within five minutes of each other. This axon showed identical results of gating currents measured in the time domain before and after the measurements. Axon held at the membrane potential of zero. The smooth curve is a theoretical fit to the 13-kHz data over a frequency range of 2 to 10 kHz for a dielectric with a distribution of time constants given by a rectangular distribution of barrier heights assuming an infinite frequency capacitance of 0.6 $\mu F/cm^2$. The dashed curve is an arc of a circle with center below the axis. The data fit this curve rather well for the parameters, appropriate to the empirical relation of Cole and Cole (1941), of $C_0 = 0.990 \times 10^{-6}\ \mu F/cm^2$; $C_\infty = 0.532 \times 10^{-6}\ \mu F/cm^2$; $\tau_0 = 1.3263 \times 10^{-5}$ ($f_0 = 12$ kHz) and $\alpha = 0.361$. The frequency labels are theoretical.

reasonable but not necessarily final. The solid curve is a fit of the 13-kHz bandwidth data to a rectangular barrier height distribution[†] (Frölich, 1958) which extends from a lower relaxation time (τ_0) of 1.95×10^{-6} sec to an upper (τ_1) of 1.01×10^{-2} sec. The central maximum of this curve thus occurs for a frequency $1/[2\pi(\tau_0\tau_1)^{1/2}]$ of 11.37 kHz. The intercept at infinite frequency of 0.6 $\mu F/cm^2$ was chosen arbitrarily. This fitted curve is meant to be illustrative only. Clearly the rectangular barrier height model does not fit well at the lower frequencies. The dashed curve is an arc of a circle with center below the axis and produces a much better fit. The data are consistent with this arc which has intercepts at a $C_0 = C_1(\omega = 0)$ of 0.990 $\mu F/cm^2$, and a $C_\infty = C_1(\omega \to \infty)$ of 0.561 $\mu F/cm^2$.

The center of the circle is depressed by 32.5° and the frequency at the center of the arc is at 12 kHz. These parameters were chosen to produce a good fit to the data.

[†] This model considers an ensemble of dipoles each with two equilibrium positions. The heights of these energy barriers are equally distributed over a certain range of barrier height.

COMPLEX CAPACITANCE AND MEMBRANE STRUCTURE

Our main interest is in trying to determine those dielectric properties of the axon membrane which are not related to ionic conduction or to the voltage-dependent saturable dielectric associated with ionic channel gating. There is also the long-term goal of relating these properties to the membrane structure. Both of these objectives are beset with numerous difficulties. Experimentally Taylor (1965) presented the first results using internal electrodes, which demonstrated that the earlier conclusions (see Cole, 1968) that the membrane was lossy were correct and did not result from the use of external electrodes where surface conductance could be a factor. He also emphasized that corrections for series and parallel resistances were crucial, particularly in regard to the interpretation of the constant phase angle which was seen on the impedance plots of the raw data (after correction for electrodes and solution impedances). Further correction for series resistance ($2.57-3.45\ \Omega\,cm^2$) presumably due to the Schwann layer and correction for an assumed leakage led to loci in the impedance plane which no longer had a constant phase angle. In the complex capacitance plane the loci were consistent with a physically reasonable model which considers a wide distribution of relaxation times for the elementary dielectric processes.

It should be pointed out that a constant phase angle in the impedance plane is quite different from a constant phase angle element which produces the circular arc with depressed center in the complex capacitance plane (Cole and Cole, 1941).

Cole and Cole (1941) reported that the capacitance (we will not speak of dielectric constants or permittivity here because we are dealing with a membrane of unknown structure) of a number of substances are represented by the empirical formula

$$C^* = C_\infty + \frac{(C_0 - C_\infty)}{1 + (j\omega\tau_0)^{1-\alpha}}$$

where τ_0 is a generalized relaxation time, C_0 is the value of C_1 at zero frequency, and C_∞ is the value of C_1 at infinite frequency. As pointed out by Cole and Cole (1941), if $\alpha=0$ this is the familiar expression for a single relaxation time Debye-type dielectric (Fröhlich, 1958) and the locus for various frequencies in the complex capacitance plane is a circle with center on the real axis. In the impedance plane the locus (see Taylor,

1965, for plots) is asymptotic at low frequencies to a real resistance of $(\tau/C_0)[1-C_\infty/C_0]$. If $\alpha \neq 0$ the locus in the complex capacitance plane is an arc of a circle with center below the real axis. As described by Cole and Cole (1941) this can be represented by an equivalent circuit consisting of a capacitance (C_∞) in parallel with a series combination of a capacitance (C_0-C_∞) and a constant phase angle element with impedance equal to $\tau(j\omega\tau)^{-\alpha}/(C_0-C_\infty)$. This does not give a constant phase angle or an arc of a circle locus in the impedance plane!

The locus of points shown in Figure 3 approaches the real axis at an angle less than 90° as does this empirical formula of Cole and Cole. In fact, the fit up to 10 kHz is quite good, including the frequency dependence. This result is quite interesting because it indicates that the distribution of relaxation times of the membrane dielectric has a very wide tail at low frequencies. We cannot say anything here about the behavior at high frequencies. Even the 70-kHz limit of the data of Taylor and Chandler (Taylor, 1965) which we discussed above is not enough to say much, as indicated by the 70-kHz frequency point indicated in Figure 3. Based on the work of Fuoss and Kirkwood (1941), Cole and Cole (1941) derived the expression for $F(s)$, the distribution of relaxation times for the depressed arc locus as

$$F(s)\,ds = \frac{\sin(\alpha\pi)\,ds}{2\pi\{\cosh[(1-\alpha)s]-\cos(\alpha\pi)\}}$$

where $s=\ln(\tau/\tau_0)$. In the arc shown in Figure 3, $\alpha=0.394$ and $\tau_0 = 1.33\times10^{-5}$ sec.

The results of Figure 3 are represented in the impedance plane in Figure 4, with and without correction for the 6.2-Ω cm^2 series resistance. We point out that this is not a constant phase angle locus and there is no obvious extrapolation to high frequency here which would produce the correct series resistance. Note that the negative of the reactance is plotted upward and the horizontal scale is 4 times the vertical.

The second objective, of relating experimental results to structure, is difficult because the membrane is clearly a mosaic and one cannot immediately relate the measured complex capacitance to the dielectric properties of a given portion of the membrane. It seems likely that the membrane is basically a lipid bilayer in which many protein complexes (probably lipo-glyco-protein complexes) are embedded. If so (Figure 1) it

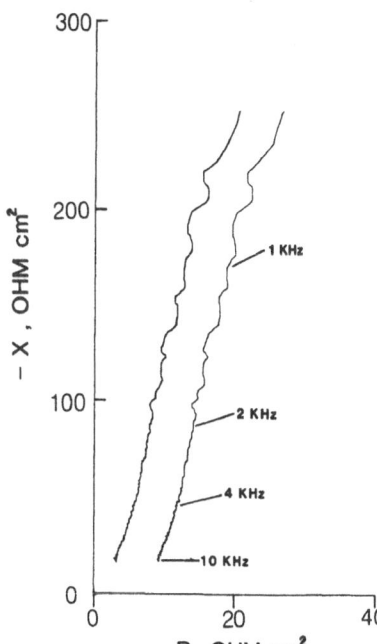

Figure 4. Data of Figure 2 plotted as impedance with (left) and without (right) the 6.2 $\Omega\,cm^2$ series resistance correction. The horizontal (real part) axis has been multiplied by a factor of 4.

is reasonable that the very-high-frequency capacitance of a good fraction of the membrane would be characteristic of a lipid bilayer. The proteins could have a high-frequency dielectric constant of 3 to 5 with uncertain thickness. A number of reports, beginning with Taylor (1965; and see Takashima, 1976, 1979; Takashima and Schwan, 1974; Takashima and Yantorno, 1977) give values of about 0.6 $\mu F/cm^2$, which seems about right. At present, it seems inescapable that the dielectric loss in the membrane is due to the penetrating proteins. We do not know of any measurements on the dielectric properties of immobilized proteins in membranes, but it would also seem reasonable that they should be similar to other charged, heavily hydrogen-bonded polymers.

ACKNOWLEDGMENTS

We would like to thank Dr. Julio Vergara for help and we acknowledge support from the following: MDA #C781030, USPHS #AM25201, and the duPont de Nemours Company.

REFERENCES

Bendat, J., and Piersol, A. (1971). *Random Data: Analysis and Measurement Procedures* (Wiley-Interscience, New York).

Binstock, L., Adelman, W. J., Jr., Senft, J. P., and Lecar, H. (1975). Determination of the resistance in series with the membrane of giant axons, *J. Membr. Biol.* **21**, 25–47.

Chandler, W. K., Fitzhugh, R., and Cole, K. S. (1962). Theoretical stability properties of a space clamped axon, *Biophys. J.* **2**, 105–127.

Clausen, C., and Fernandez, J. M. (1981). A low-cost method for rapid transfer function measurements with direct application to biological impedance analysis. *Pflügers Arch.* (in press).

Cole, K. S. (1968). *Membranes, Ions and Impulses* (University of California Press, Berkeley).

Cole, K. S. (1976). The electrical properties of the squid axon sheath, *Biophys. J.* **16**, 137–142.

Cole, K. S., and Cole, R. H. (1941). Dispersion and absorption in dielectrics. I. Alternating current characteristics, *J. Chem. Phys.* **9**, 341–351.

Cole, K. S., and Lecar, H. (1975). On the measurement of series resistance in giant axon preparations, *J. Membr. Biol.* **25**, 209–211.

Fishman, H. M., Moore, L. E., and Poussart, D. (1977). Asymmetry currents and admittance in squid axons, *Biophys. J.* **19**, 177–183.

Fröhlich, H. (1958). *Theory of Dielectrics*, second edition (Oxford University Press, London).

Fuoss, R. M., and Kirkwood, J. G. (1941). Electrical properties of solids. VIII. Dipole moments in polyvinyl chloride–diphenyl systems. *J. Am. Chem. Soc.* **63**, 385–394.

Salzberg, B. M., Bezanilla, F., and Davila, H. V. (1981). An optical determination of the series resistance in giant axons of *Loligo pealei*, *Biophys. J.* **33**, 90a.

Takashima, S. (1976). Membrane capacity of squid giant axon during hyper- and depolarizations. *J. Membr. Biol.* **27**, 21–39.

Takashima, S. (1979). Admittance change of squid axon during action potentials, *Biophys. J.* **26**, 133–142.

Takashima, S., and Schwan, H. P. (1974). Passive electrical properties of squid axon membrane. *J. Membr. Biol.* **17**, 51–68.

Takashima, S., and Yantorno, R. (1977). Investigation of voltage-dependent membrane capacity of squid giant axons, *Ann. N.Y. Acad. Sci.* **303**, 306–321.

Taylor, R. E. (1965). Impedance of the squid axon membrane, *J. Cell. Comp. Physiol.* **661** (Suppl. 2), 21–25.

Taylor, R. E., and Chandler, W. K. (1962). Effect of temperature on squid axon membrane capacity, *Biophys. Soc.* (*Abs.*) TD1.

Part II

Membrane Channels

6

Single-Channel Conductances and Models of Transport

HAROLD LECAR

Kacy Cole once said that the specific membrane capacitance is the nearest thing in biology to a physical constant. I always took that to mean that the ubiquitous $1\ \mu f/cm^2$ membrane capacitance values show that cell membranes everywhere are made of thin sheets of low-dielectric-constant material. Since a uniform low-dielectric barrier would make ion permeation much too difficult, the large conductance increases which occur during excitation are most easily explained by the activation of sparsely distributed highly conductive ion channels. One of the advances of the last few years is that we can now measure the conductance of these elementary ionic channels. I would like to first review how the unit channel conductance can be determined from electrical noise measurement and then discuss how one can directly observe the activation of

HAROLD LECAR • Laboratory of Biophysics, National Institute of Neurological and Communicative Disorders and Stroke, National Institutes of Health, Bethesda, Maryland 20205.

individual channels. At present, we see that ion-selective channels form a special class of interesting membrane proteins. There are various kinds of channel structure—not only the variety of gated channels which are switched on by various stimuli during excitation, but also cytolytic channels in various immune and attack systems, channels connecting adjacent epithelial cells at gap junctions, and probably others. The unit channel conductances form a new set of physical parameters for membrane transport which, like the capacitance, give us a bit of the picture of the membrane as a physical structure.

CHANNELS IN LIPID BILAYERS

By way of review, let me recall one set of experiments on synthetic lipid bilayers which helped to establish the channels as tangible entities. When Mueller and Rudin (1967) did their pioneering experiments on synthetic membranes, they discovered that a certain impurity, which they called EIM (excitability-inducing material) endows an inert bilayer with the kind of voltage-dependent conductance that leads to jump phenomena resembling action potentials. As with the nerve axon, excitation phenomena in the doped bilayer are concomitants of "negative conductance" Cole (1968a, b). To obtain negative conductance, there must be a range of steeply varying voltage-dependent conductances biased so that current decreases with increasing voltage. The EIM-doped bilayer provided a unique opportunity to examine the origin of the negative conductance by observing the behavior of the individual ionic channels.

Ross Bean and his co-workers (Bean *et al.*, 1969) discovered that EIM added to the membrane produces quantized steps of current, as if discrete channels or conducting domains are being created. We (Ehrenstein, Lecar, and Nossal, 1970) then proceeded to ask the question of whether EIM could be so diluted that just a small number of conducting channels might remain in steady state. With such a system, one can address an interesting question—whether each channel in an electrically excitable membrane is itself a little voltage-dependent conductance unit or whether the gating is a statistical property of the ensemble of channels. Figure 1 shows one main result of these early experiments. Under a steady voltage (voltage clamp), we see that the frequency of open–close transitions is regulated by voltage in such a way that the probability of a

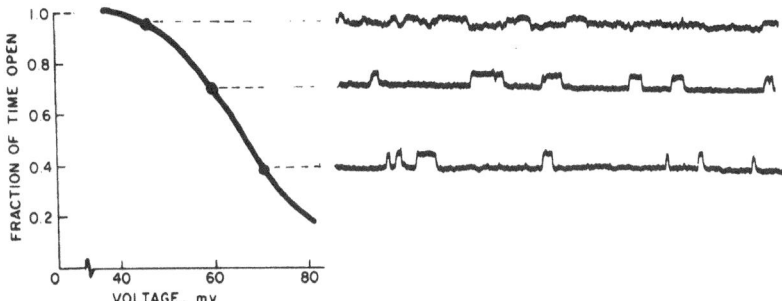

Figure 1. Explanation of EIM-induced excitability. The graph on the left shows the fraction of time a flickering channel spends in the conducting state as a function of membrane potential. The voltage dependence of the probability of being open follows the microscopically observed voltage-dependent conductance. The records at the right show typical single-channel fluctuations at three different potentials: 45, 58, and 70 mV. (The figure follows the data of Ehrenstein, Lecar, and Nossal, 1970, and Latorre, Ehrenstein, and Lecar, 1972.)

channel being in the open state as a function of voltage coincides with the macroscopic voltage-dependent conductance. Thus, for this simple system, excitability resides in gated ionic channels. There are many other ion-transporting channels with more complicated gating behavior (Ehrenstein and Lecar, 1977), but the simple picture of two-state gated channels whose transition probabilities are modulated by the stimulus has great appeal as a universal model for membrane excitation.

OBSERVING THE CHANNEL CONDUCTANCE FLUCTUATIONS IN CELL MEMBRANES

The bilayer channels illustrate how channel gating in membranes generates a unique form of electrical noise. Even in a steady state, ionic channels in a membrane can be expected to switch on and off at random. In the EIM experiment, a fortunate combination of low background noise and very large, slow channels makes detection of single-channel events easy. In living membranes, the conditions for observing single channels have been achieved only recently (Neher and Sackmann, 1976), but the method of inferring channel properties indirectly from membrane noise has been well developed over the last decade.

By now, membrane noise measurement is a large enterprise, with at least a hundred papers using noise analysis to infer properties of channels. I will not review the results of noise analysis, because there are already a number of comprehensive reviews (DeFelice, 1977; Neher and Stevens, 1976; Lecar and Sachs, 1981; Conti and Wanke, 1975). To explain how noise analysis works, let us show how the unit currents can be inferred even when the individual jumps can no longer be resolved in the noise record. This happens when the background membrane current noise is greater than the single jump current. For typical cells, this situation prevails when currents are recorded from an area of membrane greater than ~ 5 μm^2.

To see how a measurement of the total fluctuating noise power yields the unit channel conductance, imagine a membrane with N channels, each of which, when opening and closing at random, has on average a probability p of being in the open state. Then the average membrane current is

$$\bar{i} = Npi \tag{1}$$

The average power generated by the channel openings through a nominal 1-Ω resistor is given by

$$\overline{i^2} = Npi^2 \tag{2}$$

But some of this power, $N(pi)^2$, is at dc so that the average fluctuating power is just the current variance

$$\overline{i^2} - \bar{i}^2 = Np(1-p)i^2 \tag{3}$$

To infer the unit current, i, one can just take the variance-to-mean ratio,

$$\left(\overline{i^2} - \bar{i}^2\right)/(\bar{i})^2 = (1-p)i \tag{4}$$

For agonist-induced excitation, such as is produced by acetylcholine at the muscle membrane, it is usual to have $p \ll 1$, so that equation (4) gives the unit current exactly. For electrically excitable membranes, p is voltage dependent, and can often be inferred from $\bar{i}(V)/\bar{i}_{max}$. The recently developed trick of using ensemble averages (Sigworth, 1977) allows these measurements to be used for inactivating channels, such as the axon Na channel, which are not in steady state. If the channel has a fairly linear instantaneous current voltage relation,

$$i = \gamma(V - V_R)$$

one can deduce a unit conductance value, γ. Many values of γ have been obtained in this way, and tables of values are given in the reviews cited.

There was an assumption in doing this calculation, namely, that the unit current fluctuations are always rectangular jumps. This is certainly a reasonable assumption, but not necessarily the only possibility *a priori*. The final proof comes from the direct observation of the unit fluctuations.

In the simplest picture, the channel fluctuations are a Poisson process. A channel opens with uniform probability and closes without memory of the time of opening. This process, a random sequence of rectangular pulses with an exponential distribution of open-channel durations, has a characteristic frequency spectrum. The spectrum is both a useful signature for distinguishing channel noise from background and also a useful means of studying channel opening kinetics. We can explain the shape of the spectrum by considering the harmonic composition of a random train of rectangular pulses having an exponential distribution of durations.

A rectangular pulse of duration T has harmonic composition given by its Fourier transform

$$S_T(\omega) \sim \frac{\sin \omega T}{\omega T} \tag{5}$$

and consequently a random train of pulses of duration T has the spectrum

$$P_T(\omega) = |S_T(\omega)|^2 \sim \left(\frac{\sin \omega T}{\omega T} \right)^2 \tag{6}$$

If we think of each subset of pulses with durations between T and $T + dT$ as belonging to an independent Poisson process (as if someone scrambled the pulses for the convenience of this calculation), then we can get the entire spectrum by adding up the component spectra but with exponential weighting for the different pulse durations. Letting τ be the mean duration we obtain

$$P(\omega) \propto \int_0^\infty e^{-T/\tau} \frac{\sin^2 \omega T}{\omega^2 T^2} dT \qquad (7)$$

$$P(\omega) = \frac{P(0)}{1 + \omega^2 \tau^2} \qquad (8)$$

This is often called a Lorentz or Debye relaxation spectrum and is the spectrum found for the simplest type of channel noise as shown in Figure 2b. Fluctuation spectroscopy is of particular value in the study of postsynaptic channels where the fluctuation spectrum caused by a steady dose of agonist gives information about postsynaptic kinetics which cannot be obtained in other ways. Electrically excitable membranes do not give simple Poisson processes but generally exhibit spectra which can be fitted to a sum of Lorentzians.

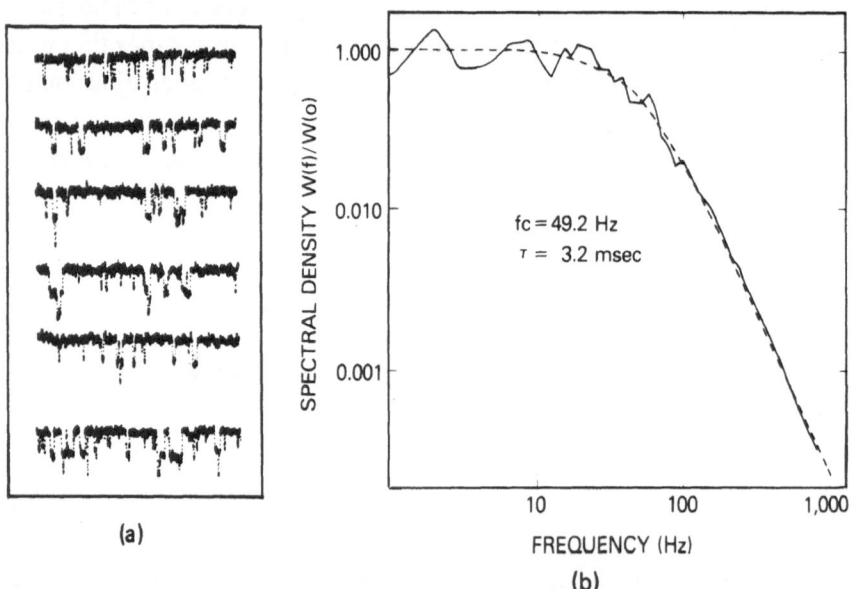

Figure 2. (a) Single-channel current jumps for postsynaptic channels excited by carbachol in tissue-cultured rat skeletal muscle. (After Jackson and Lecar, 1979.) (b) Frequency spectrum for channel fluctuations, taken by processing records such as those in (a) with a fast fourier transform computer program. (M. Jackson and H. Lecar, unpublished data.)

DIRECT OBSERVATION OF SINGLE-CHANNEL CURRENTS

Let us compare the tasks of observing the individual single EIM channels in a lipid bilayer and observing individual AcCh channels at the postsynaptic membrane. One must be able to detect the channel jump above the background. The background current noise scales with membrane area. For pure thermal noise the background current is

$$i_N = (4GkT\Delta f)^{1/2} \tag{9}$$

For EIM, $\tau \sim \sec$, $G \sim 10^{-8}$ S/cm^2, $\gamma \sim 400$ pS, and the area of the bilayer membrane in a typical experiment is ~ 1 mm^2. From noise experiments on AcCh channels we know that $\gamma \sim 40$ pS and $\tau \sim$ msec so the signal to be detected is 10^4 times smaller. In addition, $G \sim 10^{-4}$ S/cm^2, which means there is also 10^4 times more background noise. Thus to detect a single AcCh-channel current one must isolate a patch of membrane having an area ~ 1 μm^2.

The method for seeing single channels, first developed by Neher and Sakmann (1976), involves isolation of a small patch of membrane by a pipet pressed against the cell measuring the current running to ground through the pipet. To date at least 10 types of channels have been detected by the patch technique, including AcCh channels in amphibian (Neher and Sakmann, 1976), avian (Nelson and Sachs, 1979), mammalian (Jackson and Lecar, 1979), and human (Jackson et al., 1980) muscle; glutamate channels in insect muscle (Patlak et al., 1978); K channels in the squid giant axon (Conti and Neher, 1980), Na channels in muscle (Sigworth and Neher, 1980), and complement-induced channels in antibody-treated muscle (Jackson, Stephens, and Lecar, 1980). Examples of single-channel recordings taken in our laboratory are shown in Figures 2a and 3.

WHAT DO THE MEASURED γ VALUES TELL ABOUT TRANSPORT THROUGH OPEN CHANNELS?

We would like to compare the measured single-channel conductances with some theoretical limiting values. This brings up the question of what theoretical model to use for channel transport. As there has been

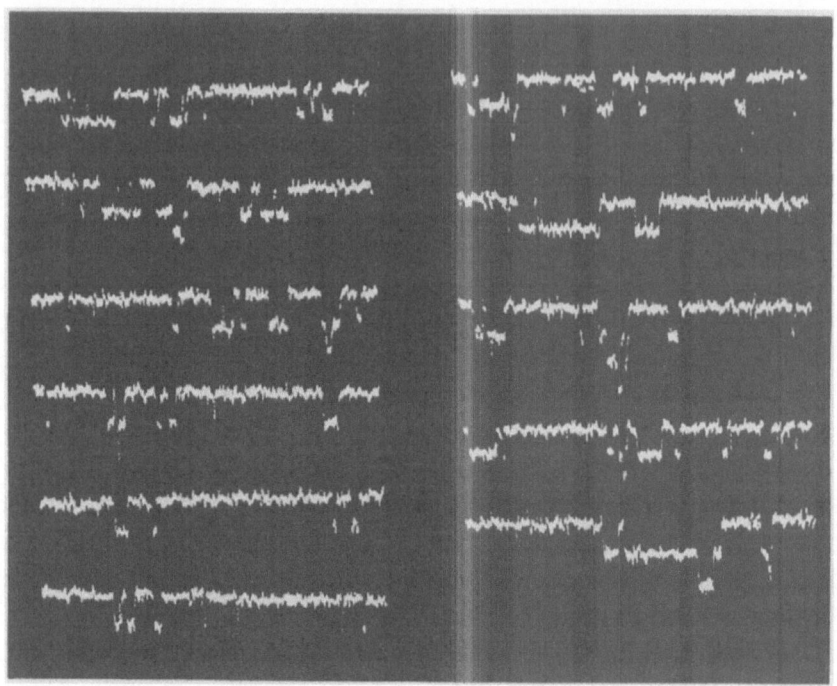

Figure 3. Single-channel jumps recorded at the postsynaptic membrane of tissue-cultured human muscle. (M. Jackson, V. Askanas, W. King Engel, and H. Lecar, unpublished data.)

an accumulation of knowledge about instantaneous $I-V$ curves and ion interaction within channels, an accurate model of channel transport should put emphasis on the constraints imposed by single-file diffusion. Even though single-file diffusion effects are inherent in transport through constricted channels, we will consider, for simplicity, a regime in which the ion concentrations are low enough for linear rate models (either in the Eyring barrier form or in the electrodiffusion form) to be applicable. That is, we assume the rate of entry of ions into the pore is sufficiently low that ion interaction or space-charge effects are unimportant. This is the approximation inherent in the independence principle (Hodgkin and Huxley, 1952) or in the constant-field electrodiffusion equation (Goldman, 1943).

As a simple paradigm, consider an ensemble of channels obeying a one-dimensional electrodiffusion equation. We wish to calculate the unit conductance for a simple water-filled pore. We might at first expect to get an overestimate for γ since ion competition for access to the pore should enter to limit the flux at higher ionic strengths.

For narrow pores, far apart, space-charge accumulation is not a factor, so the constant-field approximation can be used for the applied potential. In addition to the applied field, there is also an "image" field representing the forces required to extract the ion from solution and place it in a pore imbedded in a low-dielectric-constant membrane. This force is the gradient of a potential which we shall call $U(x)$. A recent calculation gives values of $U(x)$, expected from image-force considerations, for pores of different length and diameter (Levitt, 1978). I want to use the measured values of γ, and a simple picture of electrodiffusion through a pore represented by an effective barrier, as a means of placing limits on the barrier heights.

The current density for a membrane containing n pores is

$$J = NA_e\left[-qD\frac{dC}{dx} + \sigma\left(E + \frac{dU}{dx}\right)\right] \tag{10}$$

where A_e is the effective area of the pore, C is the concentration of the permeant ion, and D is the diffusion coefficient of ions in the pore medium. The electrical conductivity of the pore medium is given by $\sigma = mq^2C$, where m is the ionic mobility. Since D is given by the Stokes–Einstein relation, as $D = mkT$, we can replace σ by q^2CD/kT. Here q is the ionic charge, K is Boltzmann's constant and T is absolute temperature. For simplicity we can also introduce dimensionless potentials, $v = qV/kT$ and $u = qU/kT$. Equation (10) becomes

$$J = -NA_eqD\left[\frac{dC}{dx} + C\frac{d}{dx}(v+u)\right] \tag{11}$$

which can be multiplied by the integrating factor e^{v+u} to give

$$Je^{v+u} = -NA_eqD\frac{d}{dx}(Ce^{v+u}) \tag{12}$$

which can immediately be integrated to give

$$J = -NA_eqD\frac{C_2e^{v_2+u_2} - C_1e^{v_1+u_1}}{\int_0^l \exp[v(x)+u(x)]\,dx} \tag{13}$$

This is a well-known result (Hall, Mead, and Szabo, 1969). Since the potential barrier $u(x)$ goes to zero at the membrane boundaries, $u(1)=u(2)=0$. If we let $E_m=v_2-v_1$, the dimensionless membrane potential, we now have

$$J=-NA_eqD\frac{C_2e^{E_m}-C_1}{\int_0^l \exp[v(x)+u(x)]\,dx} \tag{14}$$

where l is the pore length. For a constant applied field we can write $v(x)=E_m x/l$ and then letting $\alpha=x/l$, we have

$$J(v_m)=\frac{-nA_eq^2D}{lkT}v_m\frac{C_2e^{qv_m/kT}-C_1}{\int_0^1 e^{E_m\alpha+u(\alpha)}\,d\alpha} \tag{15}$$

From this expression we can write an expression for γ measured far from the reversal potential (as is usually done),

$$\gamma=\frac{A_eq^2mC_2}{l}\left(e^{-E_m}\int_0^1 e^{E_m\alpha+u(\alpha)}\,d\alpha\right)^{-1} \tag{16}$$

The expression is simple enough; for equal concentrations and $u=0$ it would just be Ohm's law for an electrolyte-filled pore. For the slightly more general case, there is a barrier term in brackets which expresses the effect of the internal potential barriers. Another way of saying this is

$$\gamma=\gamma_0 e^{-\beta} \tag{17a}$$

where γ_0 is the ohmic conductance (roughly a lower limit) and β is a certain average potential barrier height across the pore:

$$\gamma_0=\frac{A_eq^2mC_2}{l} \tag{17b}$$

$$\beta=\ln\left\{e^{-E_m}\int_0^1\exp\left[\frac{E_m x}{l}+u(x)\right]dx\right\} \tag{17c}$$

Notice that β is an average which, because of the exponential weighting, is more sensitive to the peaks than the valleys of a barrier pattern.

For comparison, we can take experimental values of γ for channels whose dimensions are known from considerations other than single-channel conductance. For example, the dimensions of the gramicidin channel are known from molecular structure considerations, and much about the AcCh channel is known from a combination of electron microscopy and ion-permeability studies.

Table I illustrates the comparison between the lower limit (no barrier) calculated conductance and the observed conductances. From the ratio of the two, we obtain β, the average barrier in kT units. We see the barrier averages are small, of the order of $1-2$ kT, for the effective rectangular barrier. If we choose parabolic barriers, this gets corrected upward a bit to a slightly larger γ value.

Thus the main conclusion here, which incidentally also holds for a really tight channel, the Ca^{++} channel, is that the effective barriers cannot be very large. The barriers can be a bit larger than these crude estimates, for example, if $m \gg m$(water). This is a reasonable notion, since a hydrated ion going through water rips and reheals H bonds all the time, whereas motion through a channel having a polar lining of carbonyls might resemble diffusion through liquid O_2, which is less viscous than H_2O. Another possibility for some of the channels is a path which is shorter than the membrane thickness, or that the channels are funnel shaped with rate-limiting constriction over only a very narrow part of the membrane. There might be some funnel shape, but the constricted region cannot be too thin lest all the field drop across just one or two barriers, a property which leads to rather steeply rectifying $I-V$ curves. Clearly, one cannot infer too much information about the transport process from one parameter, even a parameter as fundamental as the unit conductance. A characteristic of the present work on transport through channels is the realization that there is much information—conductance as a function of ion concentration, the shapes of $I-V$ curves, and ion-selectivity patterns which must be reconciled to the emerging pictures of channel structure.

Table I. β from Measured γ[a]

Channel	Pore area ($Å^2$)	Ion concentration (mM)	γ_{max} (pS)	γ_{obs} (pS)	β (kT)
Gramicidin	13	500	105	17	1.8
AcCh	37	115	73	28	1.0
Na	24	100	21	6	1.3
Ca	15–20?	10	1.5	0.1	2.5

[a]$\beta = \ln \gamma_{max}/\gamma_{obs}$.

The more general approach to modeling channel transport is to depict the path through the channel as a series of energy barriers, chosen either to match measured I–V curves or because of some structural considerations (a number of papers on this subject are in a book edited by Stevens and Tsien, 1979). The picture discussed here, of a water-filled pore with some overall energy barrier is probably an oversimplification. The main conclusion of the discussion is that the barriers needed to fit the data are rather low.

There have been estimates of the barrier to pore-transport based on electrostatics (Parsegian, 1969; Levitt, 1978). The original motivation for the continuum electrostatics calculations was that the work needed to extract a charge from aqueous solution (dielectric constant, 80) and place it into a lipid membrane (dielectric constant, 2) is so enormous (~ 70 kT) that one could show the need for aqueous pores as the only possible way of lowering the barrier to a reasonable value. The first estimates by Parsegian (1969), for a cylindrical pore in an infinite medium, showed that a 5-Å pore could lower the energy to 12 kT. A recent calculation by Levitt (1978), in which the effect of finite membrane thickness has been properly taken into account, estimates the barrier height and shape for a pore of the dimensions of gramicidin and gives a barrier height of approximately 7 kT. If, in addition, we consider smaller electrostatic corrections such as ion interaction with polar lining of the pore (replacement of waters of hydration by carbonyl groups, for example), and dielectric shielding by the concentric protein structure of the channel, we have available terms of the order of 2–4 kT, which may bring the barrier down still further.

Thus the simple picture of a water-filled pore has not led us astray in trying to understand why channels of molecular dimensions can be such efficient ionic conductors. Such a picture does not, of course, explain anything about ion selectivity. To describe selectivity on the basis of channel structure, one must introduce explicit barriers and wells which depend on ion size.

REFERENCES

Bean, R. C., Shepherd, W. C., Chan, H., and Eichner, J. T. (1969). Discrete conductance fluctuations in lipid bilayer protein membranes, *J. Gen. Physiol.* **53**, 741–747.

Cole, K. S. (1968a). *Membranes, Ions and Impulses* (University of California Press, Berkeley).

Cole, K. S. (1968b). Membrane watching, *J. Gen. Physiol.* **51**(No. 5, pt. 2), 1s–7s.

Conti, F., and Wanke, E. (1975). Channel noise in nerve membrane and lipid bilayers, *Q. Rev. Biophys.* **8**, 451–506.

Conti, F., and Neher, E. (1980). Single channel recordings of K$^+$ currents in squid axons, *Nature* **285**, 140–143.

DeFelice, L. J. (1977). Fluctuation analysis in neurobiology, *Int. Rev. Neurobiol.* **20**, 169–208.

Ehrenstein, G., and Lecar, H. (1977). Electrically gated ionic channels in lipid bilayers, *Q. Rev. Biophys.* **10**, 1–34.

Ehrenstein, G., Lecar, H., and Nossal, R. (1970). The nature of the negative resistance in bimolecular lipid membranes containing excitability inducing material, *J. Gen. Physiol.* **55**, 119–133.

Ehrenstein, G., Blumenthal, R., Latorre, R., and Lecar, H. (1974). Kinetics of the opening and closing of individual EIM channels in a lipid bilayer, *J. Gen. Physiol.* **63**, 707–721.

Goldman, D. E. (1943). Potential, impedance and rectification in membranes, *J. Gen. Physiol.* **27**, 37–60.

Hodgkin, A. L., and Huxley, A. F. (1952). Currents carried by sodium and potassium ions through the membrane of the giant axon of *Loligo J. Physiol.* **116**, 449–472.

Jackson, M. B., and Lecar, H. (1979). Single postsynaptic currents in tissue cultured muscle, *Nature* **282**, 863–864.

Jackson, M.B., Lecar, H., Askanas, V., and Engel, W. K. (1980). Single acetylcholine channels in cultured human muscle, *Soc. Neurosci. Abstr.* **6**, 778.

Latorre, R., Ehrenstein, G., and Lecar, H. (1972). Ion transport through excitability-inducing material (EIM) channels, *J. Gen. Physiol.* **60**, 72–85.

Lecar, H., and Sachs, F. (1981). Membrane noise analysis, in *Excitable Cells in Tissue Culture*, M. Lieberman and P. G. Nelson, Eds. (Plenum, New York), pp. 137–172.

Levitt, D. G. (1978). Electrostatic calculations for an ion channel, *Biophys. J.* **22**, 209–220; 221–248.

Mathers, D. A., Jackson, M. B., Lecar, H., and Barker, J. L. (1981). Single channel currents activated by GABA, muscimol and (−) Pentobarbital in cultured mouse spinal neurons, *Biophys. J.* **33**(No. 2, pt. 2), 14a.

Mueller, P., and Rudin, D. O. (1967). Action potential phenomena in experimental bimolecular lipid membranes, *Nature* **213**, 603–604.

Neher, E., and Sakmann, B. (1976). Single-channel currents recorded from membrane of denervated frog muscle fibers. *Nature (Lond.)* **260**, 799–802.

Neher, E., and Stevens, C. F. (1977). Conductance fluctuations and ionic pores in membranes, *Ann. Rev. Biophys. Bioeng.* **6**, 345–381.

Neher, E., Sakmann, B., and Steinbach, J. H. (1978). The extracellular patch clamp: A method for resolving currents through individual open channels in biological membranes, *Pflügers Arch.* **375**, 219–228.

Parsegian, A. (1969). Energy of an ion crossing a low dielectric membrane: solutions to four relevant electrostatic problems, *Nature* **221**, 844–846.

Sigworth, F. J. (1977). Sodium channels in nerve apparently have two conductance states, *Nature* **270**, 215–267.

Sigworth, F. J., and Neher, E. (1980). Single Na$^+$ channel currents observed in cultured rat muscle cells, *Nature* **287**, 447–449.

Stephens, C. L., Jackson, M. B., and Lecar, H. (1979). Single C channel currents in living cells, *J. Immunol.* **124**, 1541.

Stevens, C. F., and Tsien, R. W., Eds. (1979). *Ion Permeation through Membrane Channels*, Vol. 3 of *Membrane Transport Processes* (Raven Press, New York).

7

Gating Kinetics of Stochastic
Single K Channels

JÜRGEN F. FOHLMEISTER
and WILLIAM J. ADELMAN, Jr.

INTRODUCTION

Recent papers by Conti and Neher (1980) and by Sigworth and Neher (1980) have shown that the conducting state of a single K^+ or Na^+ channel is "all-or-none"; that is, the channel is either open or closed. The actual occurrence of transitions between these two states appears to be a stochastic process. Nevertheless, the *probability* of finding a single channel in the open state is a function of voltage and time.

JÜRGEN F. FOHLMEISTER • Laboratory of Biophysics, Intramural Research Program NINCDS, National Institutes of Health at the Marine Biological Laboratory, Woods Hole, Massachusetts 02543, and Department of Physiology, University of Minnesota, Minneapolis, Minnesota 55455.
WILLIAM J. ADELMAN, Jr. • Laboratory of Biophysics, Intramural Research Program NINCDS, National Institutes of Health at the Marine Biological Laboratory, Woods Hole, Massachusetts 02543.

This paper concerns itself with the connection between the gating kinetics for a stochastic single channel and the gating kinetics measured for more conventionally defined conductances. In principle, it should be possible to measure directly the time course of the probability that a single channel is open by subjecting the channel many times to the same voltage clamp pulse and counting the fraction of the total number of pulses for which the channel is open at time t. A set of such voltage clamp data for different E and t could then be used to derive kinetics for changes in the probability. Such kinetics should bear a relationship to the gating kinetics for the ionic chord conductance. We therefore pose the question: is it possible to derive the gating kinetics for a stochastic single channel from the kinetics of conductance changes derived from conventional voltage clamp data?

It is shown that a quantitative description of the chord conductance alone does *not* contain sufficient information to determine even the relative probabilities that a stochastic single channel is open. However, a quantitative description of the instantaneous conductance does contain enough information for this purpose. The additional information lies in the tail currents which define the instantaneous current–voltage curves and from which the ionic driving force can be determined. Quantitative knowledge of the driving force combined with the conventional conductance kinetics is sufficient to compute the *relative* probabilities that a stochastic single channel is open. To determine the *absolute* probabilities it is only necessary to add the measurement of a single probability for a single voltage steady state.

CHORD CONDUCTANCE AND THE PROBABILITY THAT A SINGLE CHANNEL IS OPEN

Voltage clamp data typically consist of records of the membrane current which has been passed by operational amplifiers in order to generate predetermined steps in membrane potential. The measured membrane current is separated into ionic components which are then typically parametrized as the product of the two factors of "chord conductance" and the linear function $E - E_R$:

$$I_i = g_i^{\text{chord}} (E - E_R) \tag{1}$$

The temporal behavior of I_i is thus mathematically transferred to g_i^{chord} (Hodgkin and Huxley, 1952) taking care to correct for changes of the reversal potential E_K due to periaxonal space potassium accumulation for the case of the K channel (Adelman, Palti, and Senft, 1973).

An alternative parametrization, which is appropriate when considering the stochastic single-channel fluctuations, is

$$I_i = \hat{g}_i P_i f_i(E) \tag{2}$$

where P_i is the probability that a randomly chosen single channel is in the open state, $\hat{g}_i$ is a constant with the dimensions of conductance, and f_i is a voltage-dependent function which is to be determined experimentally. All gating is represented in the factor P_i. It is important to recognize that g_i^{chord} is *not* proportional to P_i unless $f_i(E) = E - E_R$. In general

$$g_i^{chord} = \hat{g}_i P_i \frac{f_i(E)}{E - E_R} \tag{3}$$

INSTANTANEOUS CONDUCTANCE AND THE DRIVING FORCE FOR ION FLUX

There are two equivalent ways to measure the function $f(E)$ of equation (2). Both involve the determination of instantaneous current–voltage curves. Such a curve is usually defined and measured for a precise history of the membrane potential which typically consists of a "long" period (seconds) at the holding potential (near the resting potential) followed by a voltage clamp pulse to a given, fixed membrane potential E for a fixed time t. Immediately following this test potential the voltage is incremented (and decremented) by various amounts ΔE and the instantaneous tail currents are measured. If the probability, P_i, that a single channel is open (or equivalently $1 - P_i$, that it is closed) does not change stepwise in response to a voltage step, then the function $f(E)$ is proportional to the instantaneous I–V curve which is defined by the instantaneous tail currents plotted as a function of membrane potential [equation (2)].

Alternatively, the *instantaneous conductance* may be measured. This quantity is defined as the slope at a particular point along an instantaneous current–voltage curve, namely,

$$g_i^{\text{inst}}(E) \equiv \lim_{\Delta E \to 0} \frac{I_i^+ - I_i^-}{\Delta E} \tag{4}$$

where I_i^- (I_i^+) is the ion current immediately preceding (following) the step ΔE. If we again assume the absence of stepwise changes in P_i, then

$$g_i^{\text{inst}}(E) = \lim_{\Delta E \to 0} \hat{g}_i P_i \frac{f_i(E^+) - f_i(E)}{\Delta E} = \hat{g}_i P_i \frac{df_i}{dE} \tag{5}$$

As for the case of chord conductance, P_i is *not* proportional to g_i^{inst} unless f_i is a linear function of E. From equations (2) and (5) we have

$$I_i = g_i^{\text{inst}} f_i \Big/ \frac{df_i}{dE} \tag{6}$$

The coefficient of g_i^{inst} is the *instantaneous driving force* at the potential E:

$$A_i(E) \equiv \frac{I_i^-}{g_i^{\text{inst}}} = f_i \Big/ \frac{df_i}{dE} \tag{7}$$

Note that only the *relative* behavior of f_i as a function of E may be derived from this expression because any constant factor, as well as P_i, cancels top and bottom.

Because the driving force should not depend on gating, the ratio of I_i^- and g_i^{inst} should remain constant even though each quantity separately will increase during a depolarizing voltage clamp pulse. Plotting I_i^- against g_i^{inst} for increasing times t should therefore result in a straight line with slope A_i. This is confirmed in Figure 1 for the potassium channel. The function $f_K(E)$ is therefore independent of the voltage history, as expected.

SPECIAL CASES OF THE FUNCTION f(E)

Consider first the possible special case in which *linear* instantaneous I–V curves are measured. The coefficient of the probability P_i in equation (2) is then the linear function $\hat{g}_i f_i = \hat{g}_i (E - E_R)$. This follows because the

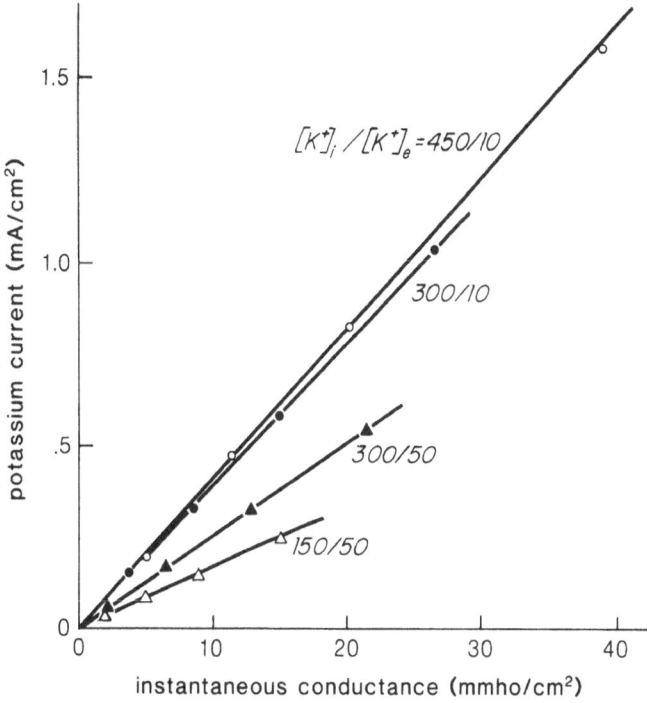

Figure 1. Potassium current plotted against instantaneous conductance for four different combinations of internal and external potassium ion concentrations. (These label the curves as ratios with all concentrations in millimolar; cf. Fohlmeister and Adelman, 1981b.) $E = -20$ mV, $T = 9°C$. Data points correspond to 1, 2, 3, and 4 msec (increasing upward diagonally) following the onset of the voltage clamp pulse from a holding potential of -60 mV.

straight line which connects the instantaneous tail currents must cross zero at the reversal potential (by definition). Therefore, $df_i/dE = 1$, $A_i(E) = E - E_R$ and comparing equations (3) and (5) leads to the conclusion that

$$g_i^{inst} = g_i^{chord} = \hat{g}_i P_i \qquad (8)$$

The gating kinetics derived for the chord conductance are then directly transferable to the gating kinetics for the relative probability of conductance, P_i, in an individual single channel. This situation may apply to the sodium channel which appears to generate linear instantaneous I–V curves (Hodgkin, Huxley, and Katz, 1952; Sigworth and Neher, 1980).

However, for the K channel the instantaneous I–V curves appear to be nonlinear and to follow closely the voltage dependence of the "constant field flux equation"

$$f_K(E) = E \frac{\exp(E_K F/RT) - \exp(EF/RT)}{1 - \exp(EF/RT)} \qquad (9)$$

(Binstock and Goldman, 1971; Begenisich, 1975). The question about the ionic flux through the open K channels (or indeed, through any channels) is *not*: Why does it follow the constant field flux equation? Rather the question is: How, and to what extent do the instantaneous I–V curves deviate from linearity? The equation (9) is merely a handy relation for the purpose of writing closed-form expressions for the stochastic single-channel rate constants once the fit of that equation has been verified experimentally.

RATE CONSTANTS FOR THE GATING KINETICS OF STOCHASTIC SINGLE K CHANNELS

To follow the Hodgkin and Huxley (1952) parametrization for the K channel, the chord conductance in the steady state is given by

$$\frac{g_K^{\text{chord}}(E)}{\bar{g}_K} = \left[\frac{\alpha_n(E)}{\alpha_n(E) + \beta_n(E)} \right]^4 \equiv n_\infty^4(E) \qquad (10)$$

Combining this equation with (3) leads to

$$P_K(E) = \frac{\bar{g}_K(E - E_K)}{\hat{g}_K f(E)} \left(\frac{\alpha_n}{\alpha_n + \beta_n} \right)^4 \equiv \left(\frac{\alpha_n^{\text{sc}}}{\alpha_n^{\text{sc}} + \beta_n^{\text{sc}}} \right)^4 \qquad (11)$$

This single expression is *not* sufficient to compute the two individual rate constants, α_n^{sc} and β_n^{sc}, for the stochastic single channel even when $f(E)$ and $\bar{g}_K/\hat{g}_K$ are known. An additional relationship between P_K and g_K^{chord} is necessary.

This additional relationship is derived from the relaxation time of the conductance which occurs between two voltage (and conductance) steady states. For the chord conductance, the relaxation times have been parametrized by Hodgkin and Huxley (1952) in terms of

$$\tau_n = \frac{1}{\alpha_n + \beta_n} \tag{12}$$

Although the *magnitudes* of $\hat{g}_K P_K$ and g_K^{chord} appear *not* to be equal [equation (3)], it can nevertheless be argued that the time interval which connects two conductance steady states is independent of knowledge of the actual magnitudes. To measure this time interval requires only that the *condition* of steady state (and not the magnitude) be recognized. Therefore, the time course for the stochastic single channel should also be equal to the time course for the chord conductance, and

$$\alpha_n^{sc} + \beta_n^{sc} = \alpha_n + \beta_n \tag{13}$$

(The equality between the relaxation times of the chord and instantaneous conductances has been demonstrated experimentally, Fohlmeister and Adelman, 1981a.) Combining equations (11) and (13) leads to the following rate constants for the gating of the stochastic single channels:

$$\alpha_n^{sc}(E) = \left[\frac{\bar{g}_K}{\hat{g}_K} \frac{E - E_K}{f(E)} \right]^{1/4} \alpha_n(E) \tag{14}$$

and

$$\beta_n^{sc}(E) = \beta_n(E) + \left\{ 1 - \left[\frac{\bar{g}_K}{\hat{g}_K} \frac{E - E_K}{f(E)} \right]^{1/4} \right\} \alpha_n(E) \tag{15}$$

These functions are plotted in Figure 2 for the ratio $\bar{g}_K / \hat{g}_K = 0.14$. This number was derived from the data of Conti and Neher (1980, Figure 2, caption, and Figure 3), from which we calculate that the fraction of time for which a single channel is open (i.e., the probability, P_K, in the steady state) is 0.036 at $E = -25$ mV. To compute the value for $\bar{g}_K / \hat{g}_K$ this value of $P_K = 0.036$ was inserted in equation (11) in conjunction with equation (9) and the rate constants $\alpha_n(-25$ mV$)$ and $\beta_n(-25$ mV$)$ from Adelman and Fitzhugh [1975, equations (13) and (14)].

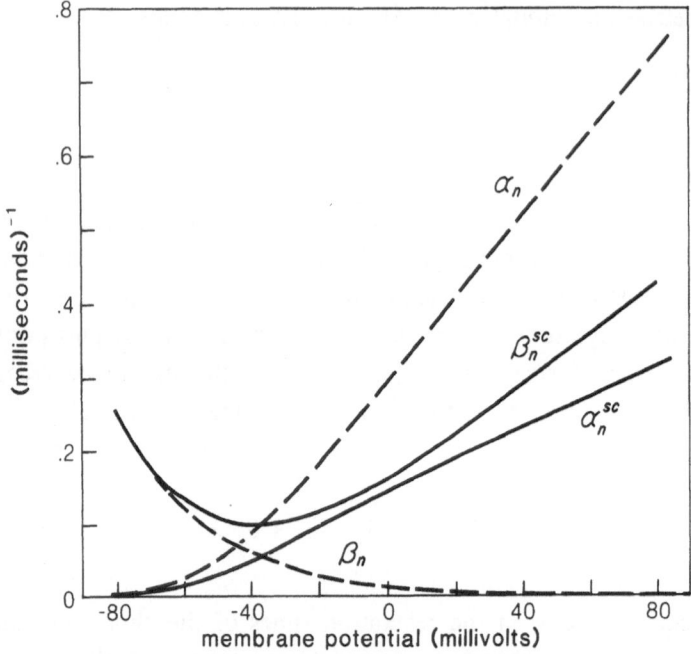

Figure 2. Rate constants as functions of membrane potential for the stochastic single K channels. Dashed curves are the rate constants for the potassium chord conductance defined in Adelman and Fitzhugh [1975, equations (13) and (14)].

THE STEADY-STATE PROBABILITIES FOR AN OPEN K CHANNEL

Figure 3 shows the relative magnitude of the chord conductance (in terms of an "n_∞^4 curve") and the *absolute* probability that a randomly chosen single channel is open, both for the steady state. The single-channel probability, P_K, appears to peak for moderate depolarizations followed by a slow decline to an asymptotic value. This is the direct consequence of two conditions which can be checked experimentally. These are (a) that the chord conductance [defined normally as in equation (1)] approaches a fixed limit for moderately large depolarizations (i.e., that a fixed value of $\bar{g}_K$ exists—at least for a normal ionic environment), and (b) that the function $f_K(E)$ deviates from linearity *in the direction of* the "constant field flux equation." If these two conditions hold simultaneously, then the single-channel probability must peak and subsequently decline as the level of steady-state potential is increased. This follows because $f(E)/(E-E_R)$ increases with increasing E, and g_K^{chord} is limited by an upper bound. These force P_K to begin a decline according to equation (3). It would therefore be desirable to recheck the behavior of the chord conductance for moderately large E in order to determine accurately its asymptotic behavior, a process that is slightly complicated by periaxonal accumulation of potassium ions.

It is perhaps tempting to suggest on "intuitive" grounds that it is unlikely that the steady-state probability for a K channel to be open should decline with increasing E, and to suggest that a careful measurement of $f_K(E)$ and g_K^{chord} would show otherwise. This is possible. However, the records of Conti and Neher (1980, Figure 2) do appear to show a decline in single-channel activity in going from $E = -8$ mV to $E = +34$ mV. If the distribution of open time intervals of a channel is independent of membrane potential, then the reduced activity implies a less frequent opening of individual channels. The data of Conti and Neher suggest an exponential distribution of open time intervals with a mean open time of 3.5 msec in the range of -35 to -19 mV. If the same distribution holds for the entire range of membrane potentials, then a declining P_K is also predicted by the data.

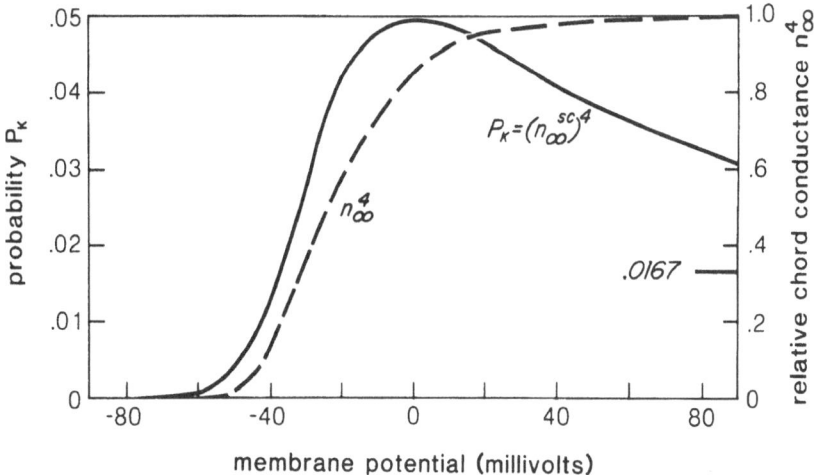

Figure 3. The probability, P_K, that a randomly chosen single K channel is open in the steady state as a function of membrane potential. Dashed curve is the relative magnitude of the potassium chord conductance in the steady state parametrized as $n_\infty^4(E)$ with $\alpha_n(E)$ and $\beta_n(E)$ from Adelman and FitzHugh (1975).

REFERENCES

Adelman, W. J., Palti, Y., and Senft, J. P. (1973). Potassium ion accumulation in a periaxonal space and its effect on the measurement of membrane potassium ion conductance. *J. Membr. Biol.* **13**, 387.

Adelman, W. J., and Fitzhugh, R. (1975). Solutions of the Hodgkin–Huxley equations modified for potassium accumulation in a periaxonal space, *Fed. Proc.* **34**, 1322–1329.

Begenisich, T. (1975). Magnitude and location of surface charges on *Myxicola* giant axons, *J. Gen. Physiol.* **66**, 47–65.

Binstock, L., and Goldman, L. (1971). Rectification of instantaneous potassium current–voltage relations in *Myxicola* giant axons, *J. Physiol.* (*London*) **217**, 517.

Conti, F., and Neher, E. (1980). Single channel recordings of K^+ currents in squid axons, *Nature* (*London*) **285**, 140–143.

Fohlmeister, J. F., and Adelman, W. J. (1981a). Anomalous potassium channel gating rates as functions of calcium and potassium ion concentrations, (in preparation).

Fohlmeister, J. F., and Adelman, W. J. (1981b). Distribution of fixed electric charge on the periaxonal and axoplasmic faces of potassium channel molecules, (in preparation).

Hodgkin, A. L., and Huxley, A. F. (1952). A quantitative description of membrane current and its application to conduction and excitation in nerve, *J. Physiol.* (*London*) **117**, 500–544.

Hodgkin, A. L., Huxley, A. F., and Katz, B. (1952). Measurement of current–voltage relations in the membrane of the giant axon of *Loligo*, *J. Physiol.* (*London*) **287**, 424.

Sigworth, F. J., and Neher, E. (1980). Single Na^+ channel currents observed in cultured rat muscle cells, *Nature* (*London*) **287**, 447–449.

Part III

Membrane Transport

8

Calculation of the Electrogenicity of the Sodium Pump System of the Squid Giant Axon

DAVID E. GOLDMAN

Ion pumps are essential constituents of cell membranes since they provide a means of compensating for ubiquitous membrane leaks and so help to maintain internal ion concentrations. In many cases these pumps generate electric current which can affect the membrane potential and are thus considered to be electrogenic. They are nearly always driven by the dephosphorylation of ATP and the process is known as active transport since it is able to transfer ions against existing electrodiffusion gradients. Thus, a mechanism driven by a biochemical reaction system has an electrophysiological result. The electrophysiological observations generated by the work of Kacy Cole have provided an important stimulus to some significant developments in membrane biochemistry.

DAVID E. GOLDMAN • Department of Physiology, Medical College of Pennsylvania, Philadelphia, Pennsylvania, and Laboratory of Biophysics, IRP, NINCDS, National Institutes of Health, at the Marine Biological Laboratory, Woods Hole, Massachusetts (permanent address).

What I have to offer here is a kind of footnote to the biophysical–biochemical dialogue. The offering is a rather simple one and can be used to obtain a quantitative treatment of electrogenesis where appropriate data are available. There have been many studies of the changes in membrane potential as an ion pump is turned on or off. A typical example, taken from the work of Gorman and Marmor (1970), is shown in Figure 1, where temperature changes have been used to control pump activity. The curve obtained at 4°C follows the constant field formulation for the resting potential fairly well whereas at 17°C the membrane is appreciably hyperpolarized with respect to the low-temperature values. There is a minimum in the curve which is centered about the region of normal external potassium. This hyperpolarization has been shown in a number of systems to be due to the presence of active transport and can be largely abolished by the application of certain metabolic inhibitors (Hodgkin and Keynes, 1955) or, where internal perfusion can be carried out (Brinley and Mullins, 1968), by removal of ATP.

In squid axons, at least, active transport fluxes appear to be independent of the membrane potential over a wide range, i.e., the relation

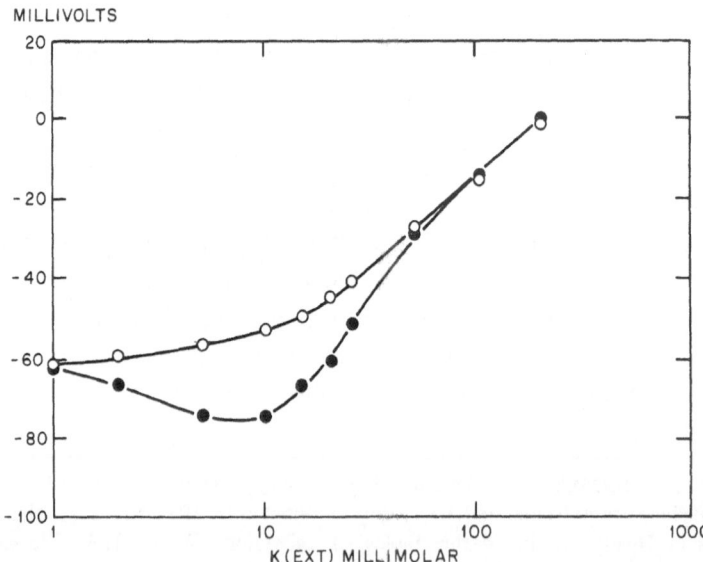

Figure 1. Resting potential of molluscan neuron. Solid circles at 17°C. Hollow circles at 4°C (redrawn from Gorman and Marmor, 1970).

between active transport and potential is not a reciprocal one. However, this is not true for all systems (Gradmann, 1978; Kishimoto, 1980).

In spite of extensive studies on the qualitative aspects of active transport in a wide variety of cells, the molecular details of the process are still not well understood. Fortunately, this knowledge is not needed here. Nevertheless, in order to make useful calculations, one must obviously have numerical data. Such data are very scarce, but thanks again to the stimulus provided by Kacy Cole there is enough information on the squid axon for my immediate purpose. Extensive measurements on internally perfused axons have been made by Brinley and Mullins (1968) and DeWeer (1970) on sodium transport and by Mullins and Brinley (1968) on potassium transport. These are shown in Figure 2.

METHOD OF CALCULATION

The process of calculating the degree of electrogenesis is as follows: The axon is considered to be in a healthy resting state with very few and

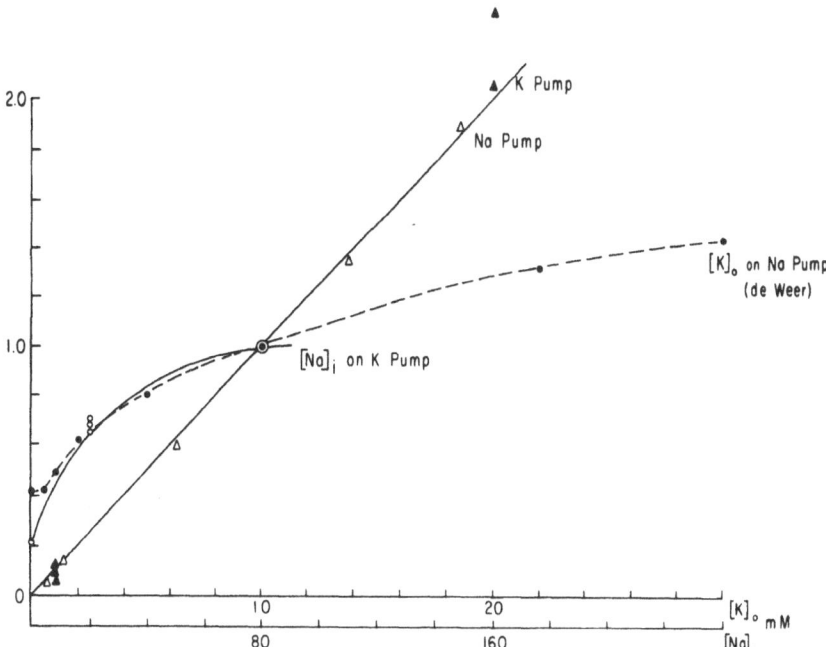

Figure 2. Sodium and potassium fluxes relative to normal state in squid axon membrane as functions of external concentrations, Mullins (1972) (reproduced by permission).

very small holes in the membrane. By a hole is meant a region in which the membrane is either absent or has lost its ion specificity and behaves like an open solution. Such holes may be as small as a few nanometers in diameter. The Nernst–Planck equations must be solved for the system but, since these are exceedingly difficult to solve exactly, I have used the constant field approximation for the intact membrane and the microscopic electroneutrality approximation of Planck for any holes which may be present, thus treating them as liquid junctions. These two elements make up the passive electric current. For the active transport I have used empirical formulas based on the data to which reference has already been made. Numerical values are as follows (these values are based on the linearity of reciprocal plots for the nonlinear curves of Figure 2):

$$I_{Na} = 9.64 \times 10^{-2} [Na]_i \left(0.24 + \frac{0.08[K]_0}{1 + 0.1[K]_0} \right)$$

$$I_K = 9.64 \times 10^{-2} [K]_0 \left(0.23 + \frac{0.042[Na]_i}{1 + 0.042[Na]_i} \right)$$

Currents are in $\mu A/cm^2$; concentrations are in mM. $[Na]_i$ is the internal sodium concentration, and, $[K]_0$ is external potassium concentration. The approximation appears to be very good; the traces of an "s" shape at the left end of one curve are too small to be considered numerically significant. On this basis I have calculated the steady-state membrane potentials, both in the presence of and in the absence of active transport. The difference between the two values is the degree of electrogenicity.

PROCEDURAL DETAILS

The calculation procedure, which involves a simple computer program, can be used for any cell system for which suitable data on active transport, ion concentrations, and ion permeabilities are available. Calcium transport can be included when it is known to be significant but at the expense of complicating the computation procedure to some extent. However, this problem does not appear to be of importance for squid axons.

Table I. Parameter Values Used in Calculations

	Ion concentrations (mM)	
	Internal	External
Na+K	350	450
Ca+Mg	—	60
Cl	70	500
Normal Na	50	—
Normal K	—	10

Other parameters
K permeability: 10^{-6}, 3×10^{-6}, 10^{-5} (cm/sec)
Permeability ratios: P_{Na}/P_K 0.04, P_{Cl}/P_K 0.12
Diffusion coefficients of ions in water (cm^2/sec):
K 9.8×10^{-6} Na 6.7×10^{-6} Cl 10.2×10^{-6}
Membrane thickness: 10^{-6} cm

Other data needed for the calculation include the composition of the environmental solutions and the permeabilities to the relevant ions. These have been taken from the literature or have been provided by sundry personal communications. The experiences and preferences of different laboratories differ somewhat. My best estimates are given in Table I. Fortunately, when accuracy is not great and biological variability is a problem, we can remember that the entire treatment is an approximation and be pleased to have useful semiquantitative results. For squid axons the electrogenic effect does not seem to be large. Alas, as things now stand, those who would like to make comparable calculations will have to find their own data for the systems in which they are interested.

The actual steps of the computation procedure are as follows:

(1) Read in the necessary parameters. I have carried out the calculations for several external–internal solution pairs.

(2) Compute the active transport currents. In this case only sodium and potassium are involved.

(3) Set the initial membrane potential at the computed constant field value for the intact membrane with no holes present.

(4) Compute the individual constant field currents including chloride.

(5) Compute the liquid junction current elements for the specified total hole area using a previously derived formula (Goldman, 1943).

(6) Obtain the total current both with and without the inclusion of the active transport.

(7) Compute the derivative of the total current with respect to the potential, i.e., the membrane slope conductance, and correct the potential using a simple Newton method.

(8) Iterate until the total current has been reduced to some preset tolerance level (.01 μamp/cm^2 has been used here). The desired steady state has then been reached and one can record the currents, the potentials, and the membrane resistance.

RESULTS OF THE CALCULATIONS

I have made calculations for several sets of external and internal sodium and potassium concentrations, potassium permeabilities, and hole areas, and I present here several plots which will serve to demonstrate the nature of the results, in particular the amount of electrogenesis. The potassium permeabilities used represent what I believe to be high, medium, and low values corresponding as they do to membrane resistances of about 300, 1000, and 3000 Ω cm^2. The shunting effect of even very small hole areas may be quite significant.

The results of the calculations are presented in several ways. Figures 3 and 4 show the active sodium and potassium currents along with the total passive current and the Na/K flux ratio as external potassium is varied. In each of these and all subsequent figures the values used for the potassium permeability and the hole area are given at the upper right-hand corner. Figures 5 through 9 show the membrane potential with and without active transport. Figures 10 and 11 show membrane resistances similarly. Figures 12 through 16 show the degree of electrogenicity as internal sodium and external potassium are varied. I should point out that when the external potassium concentration becomes large, the net active transport current changes sign and so the effect of the active transport shifts from a hyperpolarization to a depolarization. On the other hand, when the internal sodium concentration becomes large, the minimum in the potential plots becomes accentuated (see Figure 9 and compare with Figure 1). Note that the sum of the sodium and potassium concentrations on each side of the membrane is kept fixed.

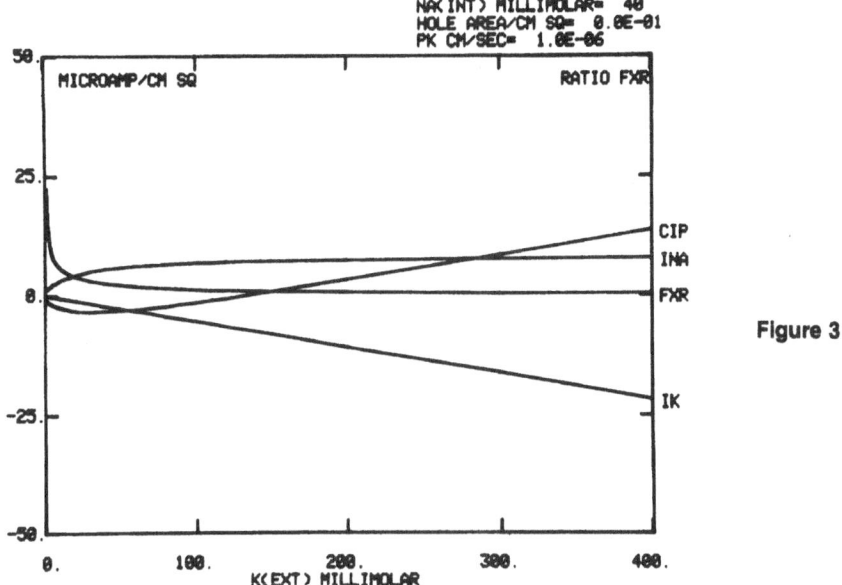

Figure 3

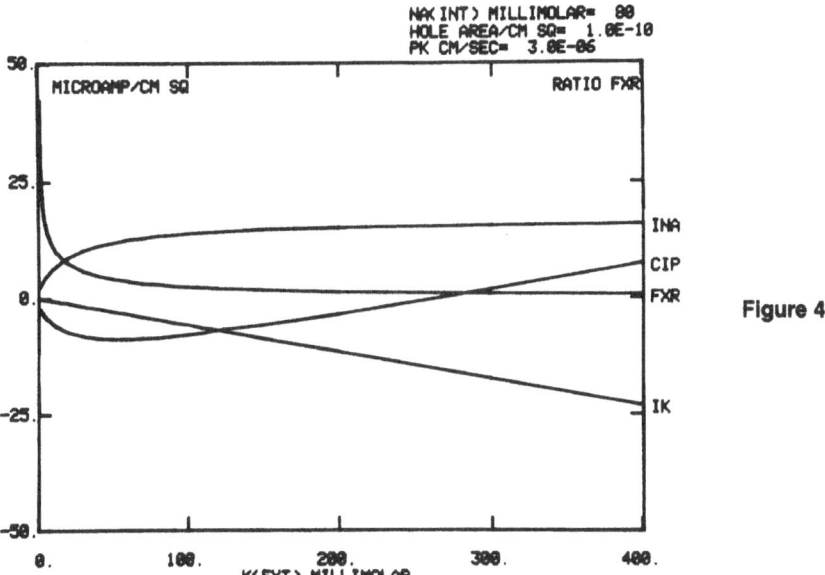

Figure 4

Figures 3 and 4. Computed current elements versus external K concentration. CIP is total passive current, INA and IK are active currents, FXR is Na/K flux ratio. Conditions at top right.

David E. Goldman

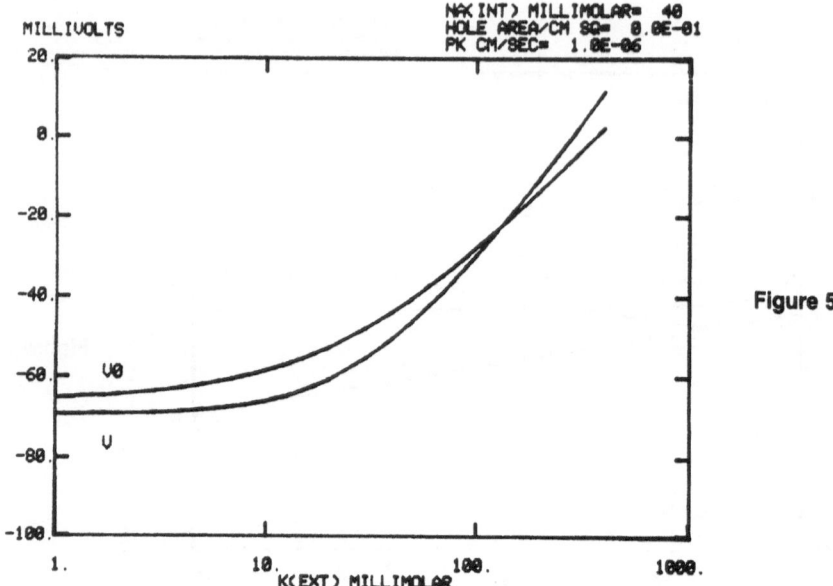

Figure 5

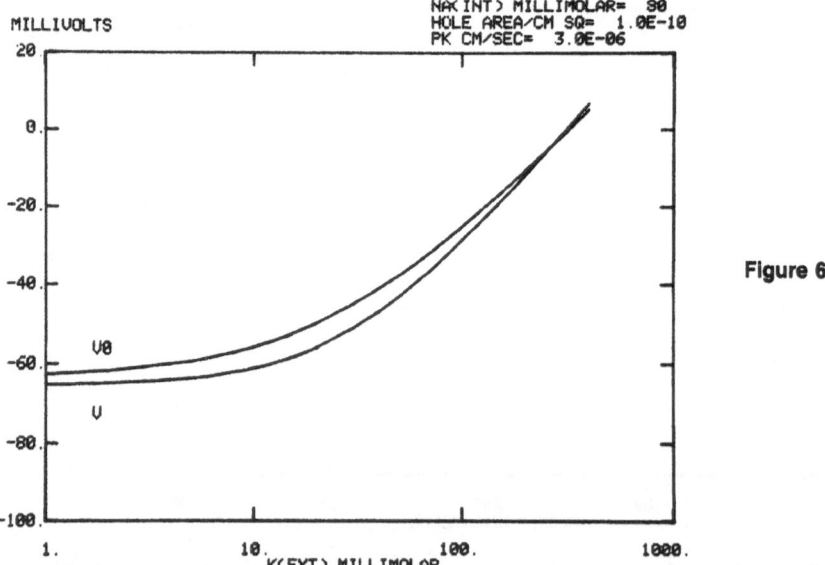

Figure 6

Figures 5 and 6. Computed resting potentials of squid axon in the presence (V) and absence (VO) of active transport. Conditions at top right.

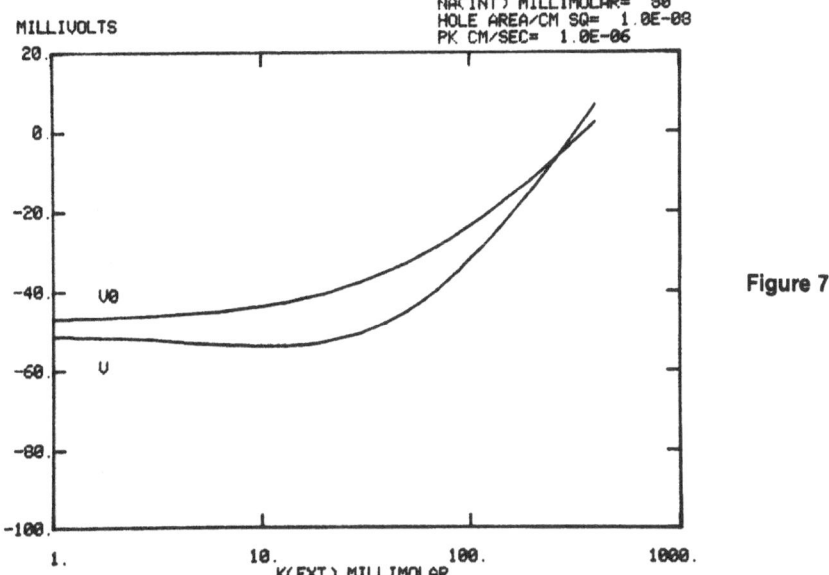

Figure 7

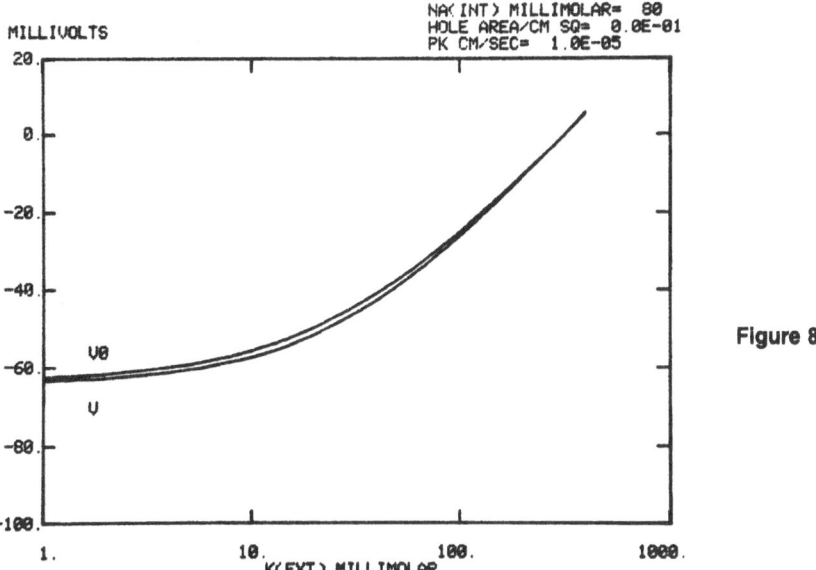

Figure 8

Figures 7 and 8. *Same as Figures 5 and 6.*

David E. Goldman

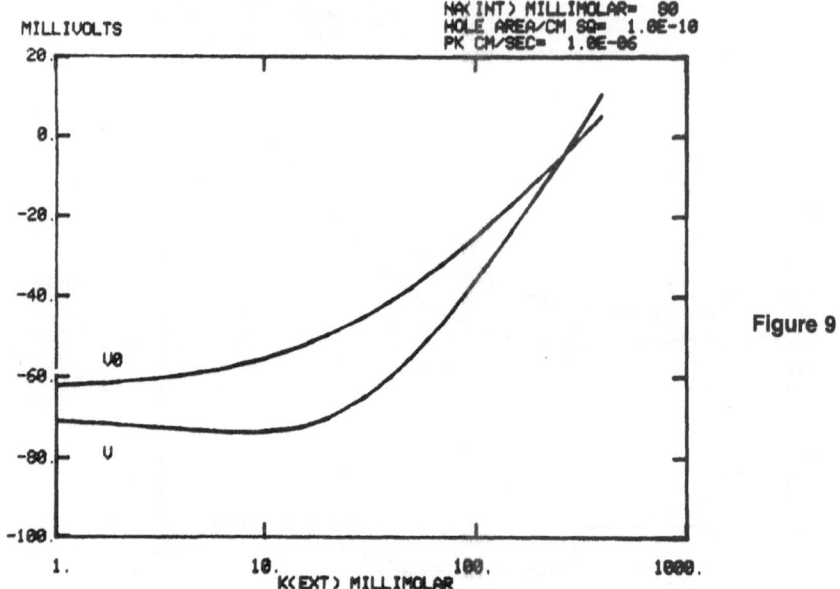

Figure 9

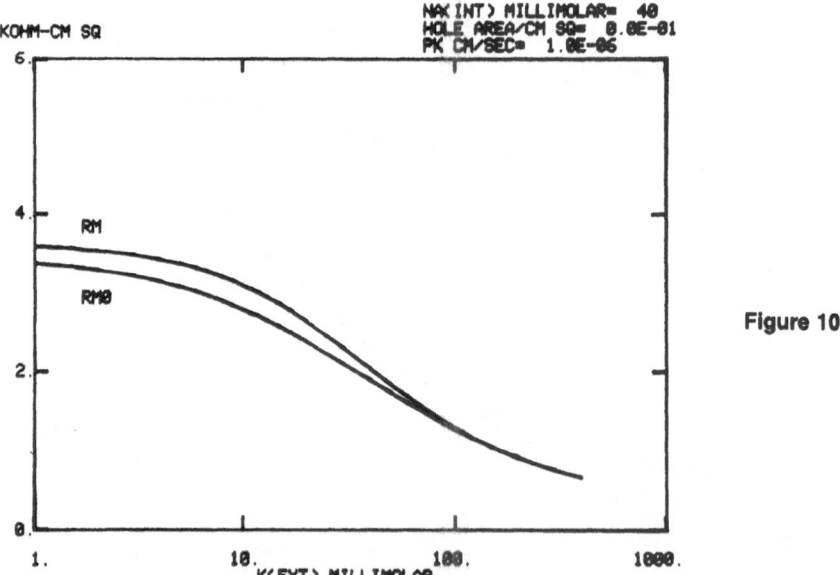

Figure 10

Figure 9. *Same as Figures 5 and 6.*
Figure 10. Computed membrane slope resistance in the presence (RM) and absence (RMO) of active transport.

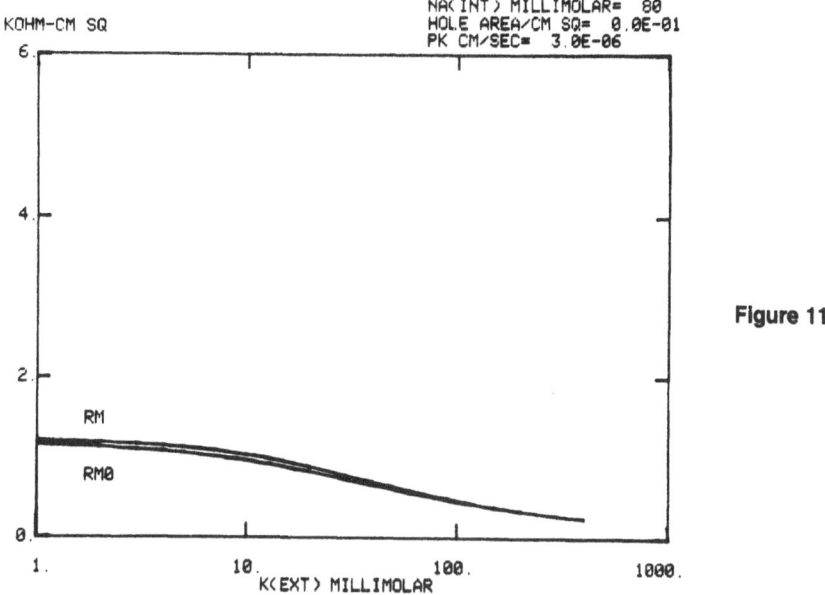

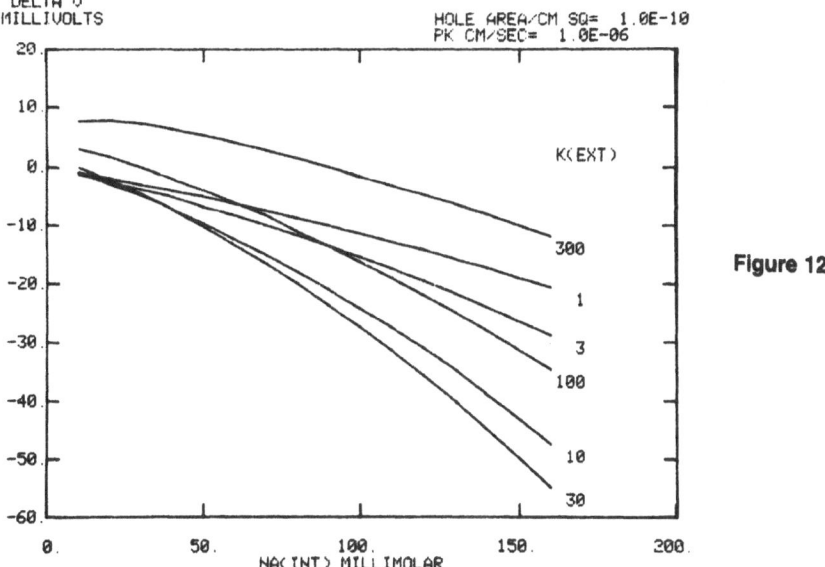

Figure 11. *Same as Figure 10.*

Figure 12. Degree of electrogenesis (DELTA V) versus internal Na concentration for several values of external K concentration.

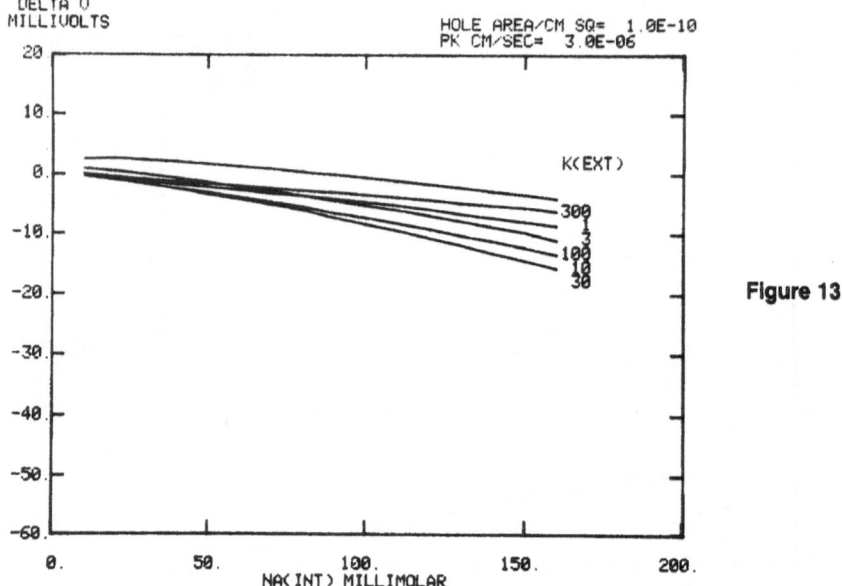

Figure 13

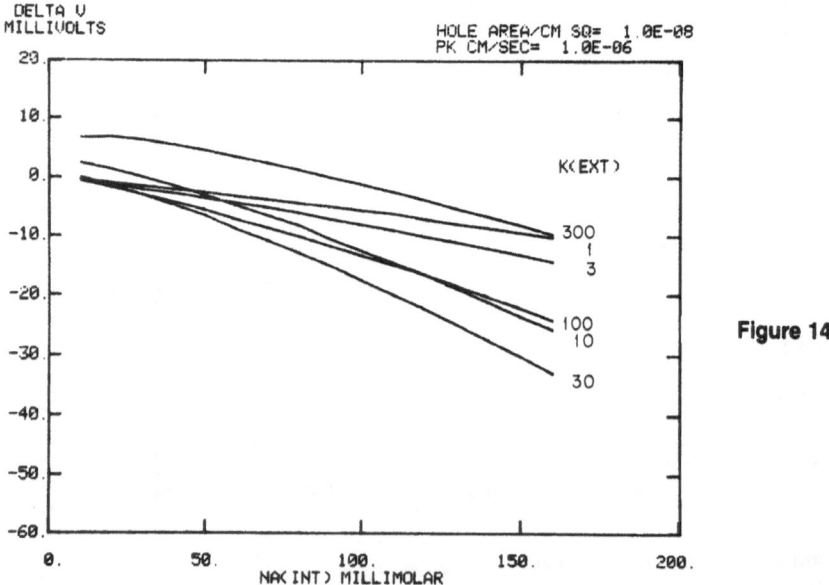

Figure 14

Figures 13 and 14. *Same as Figure 12.*

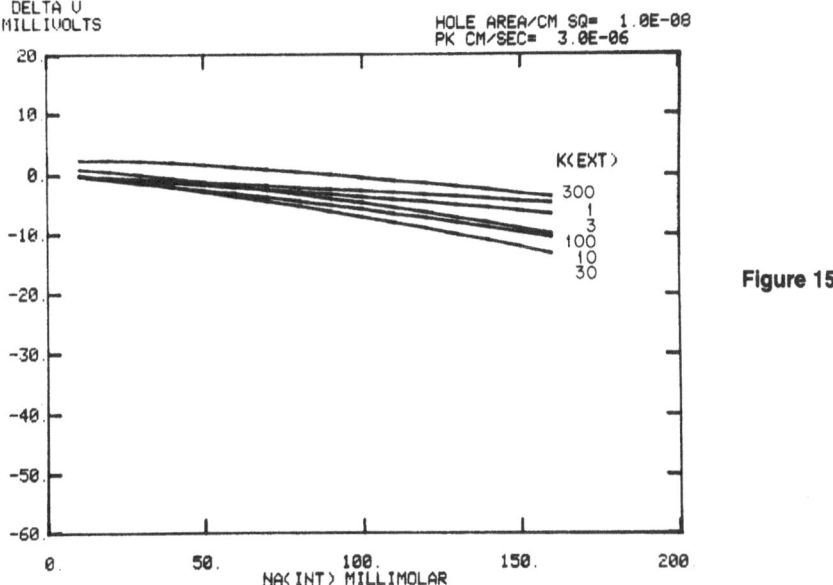

Figure 15

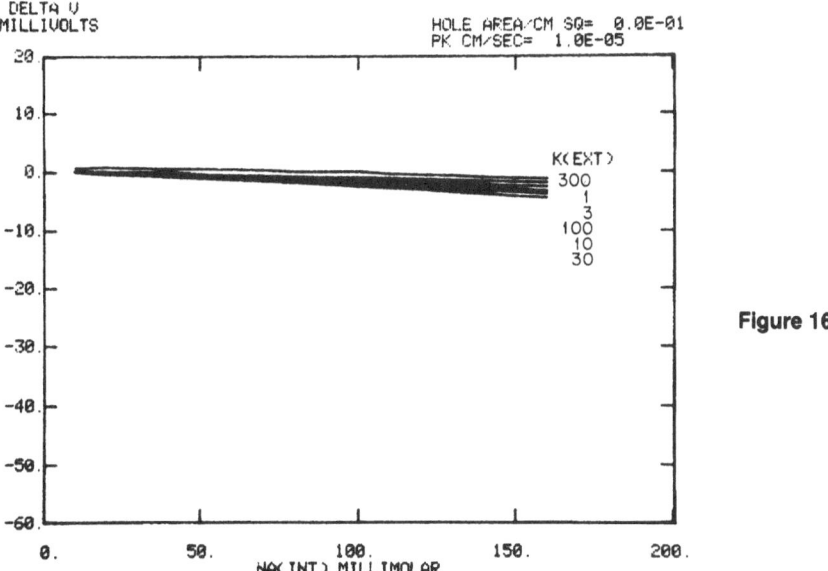

Figure 16

Figures 15 and 16. *Same as Figure 12.*

CONCLUSIONS

Examination of these figures suggests several conclusions. First, the relation of membrane potential to external potassium is similar to that shown in Figure 1 and elsewhere in the literature. The pump appears to be most effective where it is most needed for the maintenance of a resting potential able to support the generation of an action potential. Second, the membrane is able to tolerate only very small departures from complete integrity. An increase in potassium permeability of as much as a factor of 10 or the development of holes with total area of as much as 10^{-8} can quickly bring the axon perilously near to the point where the active transport system is unable to cope with the leak. Third, the quantitative aspects of the entire formulation need further development if one has a need for precise values. The calculated hyperpolarizations become quite large as internal sodium or external potassium increase substantially beyond the concentration range usually encountered in viable axons. Because of approximations in the theory and problems of experimentation, the calculated results tend to become less reliable as the currents become larger. Further progress should be possible when adequate data become available for other cell systems.

REFERENCES

Brinley, F. J., Jr., and Mullins, L. J. (1968). Sodium fluxes in internally perfused axons, *J. Gen. Physiol.* **52**, 181–211.

DeWeer, P., (1970). Effects of intercellular adenosine-5′-diphosphate and orthophosphate on the sensitivity of sodium efflux from squid axons to external sodium and potassium, *J. Gen. Physiol.* **56**, 583–620.

Goldman, D. E. (1943). Potential, impedance and rectification in membranes, *J. Gen. Physiol.* **27**, 37–60.

Gorman, A. L. F., and Marmor, M. F. (1970). Contributions of the sodium pump and ionic gradients to the membrane potential of a molluscan neurone, *J. Physiol.* (*London*) **210**, 897–917.

Gradmann, D., Hansen, U.-P., Long, W. S., Slayman, C. L., and J. Warnke (1978). Current–voltage relationships for the plasma membrane and its principal electrogenic pump in *Neurospora Crassida*. I. Steady state conditions, *J. Membr. Biol.* **39**, 333–367.

Hodgkin, A. L., and Keynes, R. D., (1955). Active transport of cations in giant axons from *Sepia* and *Loligo*, *J. Physiol.* (*London*) **128**, 28–60.

Kishimoto, U., Kami-ike, N., and Takeuchi, Y. (1980). The role of the electrogenic pump in *Chara Corallina*, *J. Membr. Biol.* **55**, 149–156.

Mullins, L. J. (1972). Active transport of sodium and potassium across the squid axon membrane, in *Symposium on the Role of Membranes in the Secretory Process*, L. Bolis, Ed. (North-Holland, Amsterdam), pp. 182–202.

Mullins, L. J., and Brinley, F. J., Jr. (1968). Potassium fluxes in dialyzed squid axons, *J. Gen. Physiol.* **53**, 704–740.

9

Depolarization and Calcium Entry

L. J. MULLINS

In presenting this lecture on the occasion of the 80th birthday of Kenneth S. Cole, I am reminded of the fact that he and I (I always think of him as K. C.) have had many conversations about the subject of what is the proper external medium with which to bathe the squid axon. Conventionally, electrophysiologists use seawater but I am sure that they recognize that this is not necessarily an ultrafiltrate of squid blood. When we first started working on the movement of calcium across the membrane of the squid axon, this question of what was the proper external calcium concentration arose once again. Now, the analysis of both seawater and squid blood yields values of about 10 mM for calcium (Shoukimas, Adelman and Sege, 1977) but there are ample reasons for supposing that neither in squid blood nor in seawater does the activity of calcium approach the value of 10 mM. This is so because in seawater

L. J. MULLINS • Department of Biophysics, University of Maryland School of Medicine, Baltimore, Maryland 21201.

there are large concentrations of sulfate, and $CaSO_4$ is a poorly dissoci-
ated substance, in addition to being rather insoluble. In blood, one has
binding to a variety of proteins and one has a rather high concentration
of sulfates, both organic and inorganic. It was for reasons such as these
that Blaustein (1974) calculated that the activity of calcium ion (Ca^{++})
was something like 4 mM around the squid axon.

The foregoing emphasizes that in talking about calcium we need to
distinguish clearly between $[Ca]_T$, the total analytical calcium measured
by atomic absorption or other modern analytical techniques, and $[Ca]$,
the ionized calcium in a phase, which can be measured with, for example,
an ion-selective electrode as a_{Ca}, the Ca^{++} activity, and when this is
divided by the ion activity coefficient, we obtain $[Ca]$. At concentrations
in the range of mM, it is possible to calculate $[Ca]$ from $[Ca]_T$ and known
dissociation constants. Intracellular values for $[Ca]$ are, however, so small
that the problem becomes more complicated. We have approached the
problem of the proper value for calcium ion concentration ($[Ca]_i$) in two
ways: first we have made measurements of the analytical calcium content
of the axoplasm extruded from squid axons under a variety of experimental conditions, and second we have measured $[Ca]_i$ with aequorin, a
protein that emits light when it interacts with calcium. Measurements of
content were made by Keynes and Lewis (1956) and these showed this to
be about 400 μmole/kg. A somewhat later study by Blaustein and
Hodgkin (1969) showed much the same value. We were, therefore, much
surprised to discover that from a freshly dissected squid axon we could
get values for $[Ca]_T$ of the order of 50 μmole/kg, or a value 8 times
smaller (Requena, Mullins, Brinley, 1979). The reason for the discrepancy was not hard to find since the axons used for analytical studies
in Plymouth have been stored in seawater for times on the order of 12 h.
When we stored Woods Hole squid axons in seawater for comparable
times, we could get comparable values for the calcium content of
axoplasm. Such an experiment strongly suggests that the concentration of
Ca^{++} in seawater is higher than that normally encountered by the squid
axon and that this excess Ca^{++} produces a net flux of calcium into the
axoplasm and hence increases the analytical content, $[Ca]_T$ there. We
have also found that these increases in calcium content of axoplasm can
be reversed by reducing $[Ca]_o$ to values to the order of 3 mM. The
calcium content of axoplasm could, in the course of an hour, be brought
down to values substantially smaller than it was initially, and in the

course of two hours, one could easily reach the initial values that fresh axoplasm has for calcium content. This is reassuring because it suggests that no irreversible damage has been done to the machinery which regulates the calcium of nerve and that calcium content can rise and fall as the calcium concentration in the bathing fluid goes up or down.

It is not, however, the analytical $[Ca]_T$ that is of most interest in a nerve fiber, but it is rather the $[Ca]_i$ that is of the greatest interest to physiologists. There are relatively fewer measurements of this quantity because it is, in fact, difficult to measure. We have (DiPolo *et al.*, 1976) confined the light-emitting protein, aequorin, inside a dialysis capillary at the center of an axon. Here the protein equilibrates with the [Ca] of the axoplasm and with suitable *in vivo* calibration it is possible to specify that the internal calcium ion concentration $[Ca]_i$ is about 30 nM. These experiments with aequorin are very useful in demonstrating the fact that ionized calcium rises if $[Ca]_o$ is higher than a certain value, and falls if $[Ca]_o$ is below it. A record of these aequorin studies is shown in Figure 1 and indicates that a change from 10 mM [Ca] seawater to one containing only 3 mM changes the slope of the aequorin glow versus time from one that was rising to one that is now virtually flat. The result of a number of studies similar to those shown in Figure 1 suggests that the balancing concentration $[Ca]_o$ is of the order of 2.8 mM. Thus, both analytical studies and ionized calcium measurements agree in showing that this concentration in seawater that is sulfate free is one where the axon neither gains nor loses calcium.

CALCIUM ENTRY WITH STIMULATION

The findings described above are of rather recent date, but the story of calcium entry into squid axons really starts with a paper by Hodgkin and Keynes (1957) where they showed that stimulating squid axons in a solution of ^{45}Ca caused an entry of calcium into the axoplasm which was directly proportional to the concentration of calcium in seawater bathing the axon. Hodgkin and Keynes also showed that there was an increase in the resting entry of calcium into fibers that was linear with $[Ca]_o$ so that it appeared that a diffusion pathway was utilized for the movement of the calcium; however, the entry was by any standard extremely small. Their findings were in fact that one could have an entry of 10^{-15} mole

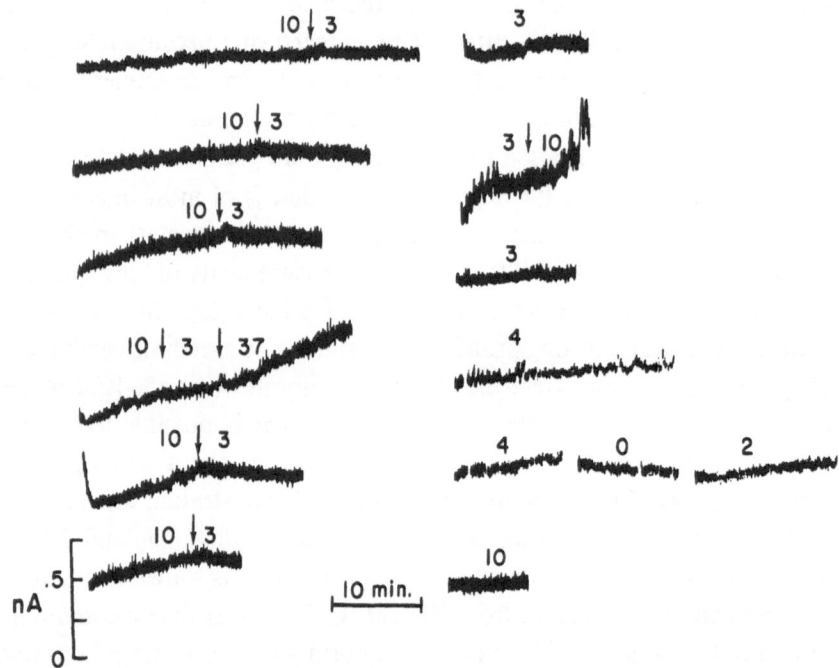

Figure 1. For a series of axons, the light output from aequorin (ordinate) as a function of time (abscissa) when the [Ca] of seawater is changed in the range of 0–10 or 37 mM. The most common transition was from 10 to 3 mM and this most commonly leads to light emission that is independent of time. By contrast, a change from 3 mM Ca to 10 or 37 mM Ca invariably leads to an increase in light emission (hence in [Ca]$_i$) with time. (Reproduced from DiPolo et al., 1976.)

Ca/cm^2 per impulse per mM of external calcium. That is, one femtomole of calcium per square centimeter entered for each millimole per liter of Ca present in seawater.

Subsequently, Baker, Hodgkin, and Ridgway (1971) were able to show, using aequorin as an indicator of calcium entry, that under voltage clamp one could divide calcium entry with depolarization into a calcium entry sensitive to external tetrodotoxin (TTX) that was presumably via the sodium channels, and a TTX-insensitive calcium entry that was presumably via another mechanism. A detailed study of this calcium entry under voltage clamp conditions showed that for rather short pulses (times of the order of an action potential) almost all the entry was TTX-sensitive, whereas for long pulses a negligible fraction was TTX-sensitive. Since it has been known for some time that in the nerve terminal there is a calcium entry that is associated with the release of transmitter, a reasonable supposition was that these so-called "late" calcium channels were not only distributed in the region of the nerve terminals, but occurred, albeit much more sparsely, all along the axon. Another finding of Baker, Hodgkin, and Ridgway was that while axons often responded to stimulation with light output that was linear with the frequency of stimulation, in some exceptional axons this response could be as high as the 2.4 power of the stimulation of frequency. Such "sensitive" axons did not have a higher resting glow (indicating that $[Ca]_i$ was normal) but did behave as though much more calcium were entering per bioelectric event than was normal. When the responses to a voltage clamp as a function of clamp potential were analyzed in detail, it was clear that calcium entry was maximal when the absolute value of the membrane potential was about zero, and entry declined as this potential became more inside positive. This response was unusual since the equilibrium potential for calcium is a large positive number ($+145$ mV). Thus it appears necessary to suppose that the permeability of the channel to calcium is declining as the membrane is depolarized from zero to inside positive membrane potentials.

A study rather similar to that of Baker, Hodgkin, and Ridgway was one carried out by Rojas and Taylor (1975); however, they used ^{45}Ca as an indicator of calcium entry and perfused squid axons as a vehicle for the studies. Both the resting influx of calcium and that component of calcium entry that was a "late" one were very much smaller than the expected values for calcium entry measured in intact axons. Recall that these internally perfused axons had CsF as an internal medium and hence were substantially different from normal axons. The axons did, however, have the normal entry of calcium via sodium channels.

One of the obvious explanations for a TTX-insensitive calcium entry would be that calcium was going through potassium rather than sodium channels. Since potassium channels do not have inhibitors of the specific sort such as TTX, it was a matter of some experimental difficulty to try and rule out the possibility that calcium movement actually did take place through potassium channels. Use was made of the divalent cation Mn^{++} which inhibited late calcium entry and to argue that it did not affect potassium channels as much as it did the late current. Substances such as TEA are in general unsatisfactory because they only inhibit potassium movement in an outward direction, and one is expecting a calcium entry via movement inward through potassium channels if indeed such a movement exists. Recently (Keynes *et al.*, 1979, and personal communication) evidence has been obtained showing that calcium does move through potassium channels in internally dialyzed squid axons and the effect is especially noticeable when after a depolarizing pulse the membrane is returned to normal polarization and the inward tail current through the potassium channels is highly sensitive to the presence or absence of calcium in the seawater. We have, therefore, the conclusion that calcium can move inward not only by diffusion, and by moving through sodium channels, but also through "late" channels and through potassium channels.

CALCIUM ENTRY WITH STEADY DEPOLARIZATION

If instead of repetitive stimulating pulses, a squid axon is held depolarized by increasing the [K] in seawater to for example 100 mM, then as Baker, Meves, Ridgway (1973a, b) have shown, there is a phasic entry of calcium as judged by aequorin light emission changes where the light rises rapidly to a peak and then declines to a steady value which is higher than the initial resting value for light emission. A detailed analysis of this effect showed that there was no sensitivity to TTX treatment, implying that Na channels did not carry any appreciable fraction of the calcium that entered during these maintained depolarizations, but that the effect could be effectively abolished by applying agents such as Mn^{++}. Since these properties of differential sensitivity to Mn^{++} and to TTX are similar to those described above for responses to short and long pulses, the observations were interpreted by the authors of the paper to indicate that again one was dealing with the "late" calcium channel. The channel appeared to have the following characteristics: it inactivated since testing with solutions of elevated [K] concentration spaced closer in time than about 5 min apart resulted in the second test depolarization giving a smaller response than the first; total recovery from inactivation appeared to take on the order of 10 min. The same sorts of results could be obtained if the depolarization were brought about by electrical currents rather than by elevating the [K] of seawater, so that the response appeared to be genuinely the result of the depolarization and not that resulting from some chemical property of potassium. The final part of the observations with steady depolarizations was that there was always a continued elevated calcium entry over control levels in depolarized axons in addition to the phasic response which inactivated. Thus, it was possible that there were two sorts of processes going on simultaneously in the measurements that had been made.

CALCIUM ENTRY BY Na–Ca EXCHANGE

The original observations of Baker *et al.* (1969) showed that squid axons bathed in Na-free solutions showed a greatly enhanced uptake of calcium as judged by ^{45}Ca influx measurements. Furthermore, this calcium entry appeared to be highly dependent upon $[Na]_i$ so that the reasonable interpretation placed upon these findings was that sodium-free solutions, by reversing the sodium gradient across the membrane, caused the uptake of calcium rather than its extrusion. Measurements of calcium efflux in dialyzed axons (Mullins and Brinley, 1975) showed that hyper-polarizing the membrane increased calcium efflux and depolarizing it had the opposite effect. These findings suggested that the Na–Ca transfer mechanism might be electrogenic in nature and move more sodium charges in one direction than calcium charges in the opposite direction. If this is so, then one would expect that membrane potential itself would have an influence on the rate of movement of the Na–Ca exchange mechanism. A preliminary (and unpublished) measurement of the sodium efflux from squid axons that is coupled to calcium entry is shown in Figure 2. Here it is clear that depolarization enhances the sodium efflux that is dependent on $[Ca]_o$, again confirming the notion that there must be a component of calcium influx that is both voltage sensitive and mediated by Na–Ca exchange. These measurements confirm those of Baker and McNaughton (1976).

It may be helpful at this point to recall that the treatment presented so far has described, in addition to passive entry, the following voltage-dependent modes of calcium entry during depolarization: (1) sodium channels, (2) potassium channels, (3) "late" channels, and (4) Na–Ca exchange operating in a reverse mode. Aside from the finding that at least for the duration of a normal action potential in squid axons most of the calcium entry is by sodium channels, there is little information in the literature that would enable one to relate the quantities of calcium that might move via the various mechanisms listed above. One way of being able to understand the relationships between calcium entry via this array

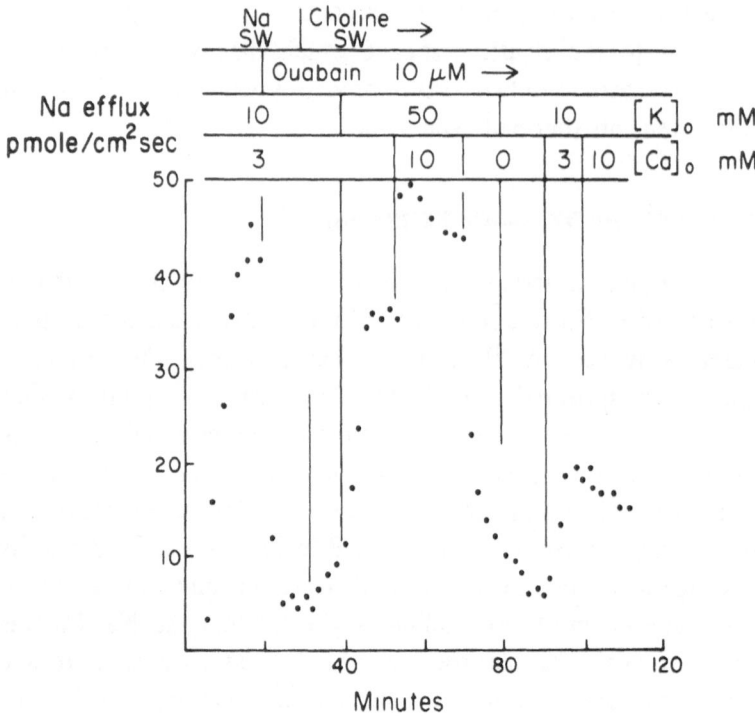

Figure 2. A squid axon injected with ^{22}Na has its Na efflux (ordinate) shown as a function of time. The axon was stimulated for 30 min in K-free seawater to increase $[Na]_i$ before injecting isotope. The first step in this study was to poison Na efflux with ouabain and a second step was to make the axon Na-free to enhance Ca entry in exchange for Na_i. Depolarization with 50 mM K then led to a large increase in Na efflux that was Ca_0 sensitive. At normal 10 mM K the Na efflux less that in 0 was ~10 pmole while with depolarization this increased to 50 pmole or a net change of 30 pmole, which for 4 Na/Ca would be a Ca influx of 7 pmole/cm^{2}sec. (Unpublished data, Mullins and Brinley.)

of mechanisms is to understand more about Na–Ca exchange itself and in particular to understand its voltage sensitivity.

If a carrier translocated two sodium per calcium across the membrane (i.e., an electroneutral arrangement), then calcium movement via this mechanism would not be voltage dependent. It would also result in $[Ca]_i$ being 1/100th of $[Ca]_o$ since for this arrangement $[Ca]_o/[Ca]_i$ equals $[Na]_o^2/[Na]_i^2$ and the sodium ratio across the squid axon membrane is about 10. This value for $[Ca]_i$ is more than one thousand times higher than experimentally measured values, so that clearly if Na–Ca exchange is maintaining the low $[Ca]_i$ of axoplasm, more sodium must move in one direction for calcium moving in the opposite one. The mechanism is electrogenic if the coupling ratio r equals three or more and there is obviously a wide choice of coupling ratios. The second property, however, namely, the voltage sensitivity for calcium entry is also involved here. The change in calcium efflux with membrane potential appears to be e-fold per 25 mV change in membrane potential and there is a similar voltage sensitivity for calcium influx, hence the flux ratio which is calcium efflux divided by calcium influx is proportional to $\exp(-2EF/RT)$, an expression appropriate for the movement of two net charges per cycle of the carrier. Thus, measurements of this sort suggest that four sodium ions might move per calcium ion translocated rather than three; this also allows the ultimate $[Ca]_i$ to be in the low nM region or in a range where measurements indicate that $[Ca]_i$ actually is. The proper expression then for the free energy available from the sodium gradient to drive transport is $4F(E-E_{Na})$ and the energy required by calcium for its transport is given by $2F(E-E_{Ca})$, where F is the Faraday constant, E is the membrane potential, and E_{Na} and E_{Ca} are the respective equilibrium potentials. These relations can be reduced to the equation $E=2E_{Na}-E_{Ca}$. With the usual value of $E_{Na}=+60$ mV and $E_{Ca}=+145$ mV, the value of E, which we may call the reversal potential E_R of the carrier current, is -25 mV.

Without detailed model considerations we have no basis for assuming how calcium current might vary with membrane depolarization, but at least for small depolarizations it might be expected to increase exponentially as one displaces the membrane potential from its resting value. This means that we should expect a calcium entry into squid axons to start at -25 mV and to increase rapidly with depolarization until some other saturating process sets in. It also means that calcium entry by this mode should be critically dependent on $[Na]_i$. This is so because Na–Ca exchange is a coupled movement of sodium in one direction and calcium in the other, and in fact it can be shown that for the simplest possible cases there should be a fourth-power dependence of calcium entry on $[Na]_i$. We have, therefore, carried out some steady depolarization experiments entirely analogous to those of Baker, Meves, and Ridgway (1973a, b). One such experiment is shown in Figure 3 (from Mullins and Requena, 1981) where we depolarized an aequorin-injected squid axon with 200 mM potassium and 50 mM calcium seawater and obtained a result shown in the leftmost trace of Figure 3, the usual phasic

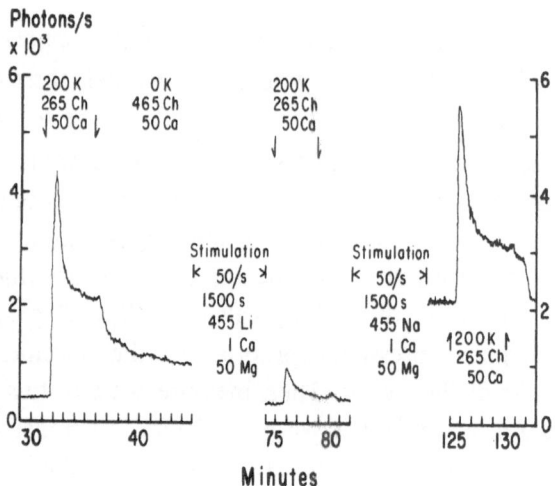

Figure 3. An aequorin-injected squid axon kept in 50 mM Ca, choline seawater responds to a change to 200 mM K, choline seawater with the response shown on the left. Time, shown on the abscissa, is measured from the time of aequorin injection. After the response was produced, the axon was stimulated in Li$^+$ seawater to reduce [Na]$_i$ and then retested with 200 mM K yielding the response shown in the middle trace. Finally, a second period of stimulation, but this time in Na-containing seawater, thus restoring [Na]$_i$ to about normal values, yielded the response shown on the right. (Mullins and Requena, 1981.)

calcium entry followed by a plateau. Removal of the high potassium solution led to a recovery of the base line toward its initial value. After this test, the axon was placed in a lithium-containing seawater with low calcium and stimulated for 30 min at 50/sec. This treatment has been shown by other straightforward analytical measurements to reduce the [Na]$_i$ of axoplasm to almost half its former value since lithium enters through the sodium channels during stimulation and the sodium inside the fiber exits through the same channels as stimulation proceeds. A retest of this axon after reducing [Na]$_i$ gave the response shown in the middle trace of Figure 3, while a final stimulation in sodium-containing seawater which approximately brought back the original [Na]$_i$ led to a recovery of the sensitivity of the axon to depolarization. A conclusion from experiments of this sort is therefore that all or essentially all of the calcium entry in response to steady depolarization in squid axons is a result of the operation of the Na–Ca exchange system in the axon, which operates in a direction that is dependent upon the direction of the sodium electrochemical gradient. If the membrane potential is reduced to values below the reversal potential for the carrier already discussed above, then the direction of net movement of calcium is also reversed from outward to inward; similarly, if the sodium electrochemical gradient is reduced by changing the concentration of sodium on the outside of the membrane to a value lower than its concentration inside, the direction of calcium movement is similarly changed from a net outward to a net inward movement. Note that what is being affirmed here is that the sodium electrochemical gradient consists of two terms, one in chemical potential energy and another an electrical term, and these two terms may be manipulated independently in order to produce net calcium movements in one direction or in the other.

Similar measurements using repetitive electrical stimulation of the axon rather than steady depolarization show that for experiments (Figure 3) where $[Na]_i$ was substantially reduced, the response to stimulation was not affected, and hence these measurements confirm an earlier conclusion, which was that for depolarizations of times of the order of an action potential most of the calcium entry is via the sodium channels and this cannot be expected to be influenced by modest changes in $[Na]_i$. If it were influenced one might expect a greater sodium entry as $[Na]_i$ is reduced. The experiments reported above (Mullins and Requena, 1981) provide no support for the idea that there is a "late" calcium channel in squid axons since the response to steady depolarization can be virtually abolished when $[Na]_i$ is made low enough. The remaining mode of calcium entry (that of calcium movement through the potassium channels) has not been sufficiently well studied to make it possible to characterize this mode of calcium entry very much. It presumably was involved in the experiments of Rojas and Taylor (1975) since their internal medium contained no sodium, and yet they found a small but definite calcium entry with depolarization that was TTX insensitive.

It is reasonable to conclude that for times of the order of an action potential, depolarization leads to an entry of calcium via both sodium and potassium channels. If the time frame for depolarization is not 1 msec (the duration of an action potential) but rather 1 min, then virtually all of the calcium entry is not by a channel mechanism, but by Na–Ca exchange running backward whereby it introduces calcium in exchange for internal sodium. The reason why channels are unimportant in long-lasting depolarizations is that both sodium and potassium conductances inactivate, and while the potassium conductance is far slower in activating, this slowness can be described in terms of hundreds of milliseconds whereas the time period under consideration is minutes. Most bioelectric activity operates on a millisecond time scale so that calcium entry from Na–Ca exchange may not in general be an important physiological mechanism. There is, however, an exception, namely, that of the cardiac action potential, whose duration can be many hundreds of milliseconds. It has been suggested (Mullins, 1979) that for this long duration action potential, calcium entry via Na–Ca exchange is an important contributor to the contraction of cardiac muscle. The experiments reported above would bear out this contention. I would close by reminding electrophysiologists that what is being described by Na–Ca exchange is a carrier-mediated process that not only transfers significant quantities of calcium across the cell membrane, but produces a current that is oppositely directed to that of calcium movement, hence one that could lead to a novel bioelectric effect whereby one has the entry of calcium as part of an outward net current.

REFERENCES

Baker, P. F., Blaustein, M. P., Hodgkin, A. L., and Steinhardt, R. A. (1969). The influence of calcium on sodium efflux in squid axons, *J. Physiol.* **200**, 431–458.

Baker, P. F., Hodgkin, A. L., and Ridgway, E. G. (1971). Depolarization and calcium entry in squid axons, *J. Physiol.* **218**, 709–755.

Baker, P. F., and McNaughton, P. A. (1976). The effect of membrane potential on the calcium transport systems in squid axons, *J. Physiol.* **260**, 24–25P.

Baker, P. F., Meves, H., and Ridgway, E. G. (1973a). Effects of manganese and other agents on the calcium uptake that follows depolarization of squid axons, *J. Physiol.* **231**, 511–526.

Baker, P. F., Meves, H., and Ridgway, E. G. (1973b). Calcium entry in response to maintained depolarization of squid axons, *J. Physiol.* **231**, 527–548.

Blaustein, M. P. (1974). The interrelationship between sodium and calcium fluxes across cell membranes, *Rev. Physiol. Biochem. Pharmacol.* **70**, 33–82.

Blaustein, M. P., and Hodgkin, A. L. (1969). The effect of cyanide on the efflux of calcium from squid axons, *J. Physiol.* **200**, 497–527.

DiPolo, R., Requena, J., Brinley, F. J., Jr., Mullins, L. J., Scarpa, A., and Tiffert, T. (1976). Ionized calcium concentrations in squid axons, *J. Gen. Physiol.* **67**, 433–467.

Hodgkin, A. L., and Keynes, R. D. (1957). Movements of labelled calcium in squid axons, *J. Physiol.* **138**, 253–281.

Keynes, R. D., and Lewis, P. R. (1956). The intracellular calcium contents of some vertebrate nerves, *J. Physiol.* **134**, 399–406.

Keynes, R. D., Malachowski, G. C., and Van Helden, D. (1979). Recording of ionic and gating currents in giant axons under computer control, *J. Physiol.* **287**, 1P.

Mullins, L. J. (1979). The generation of electric currents in cardiac fibers by Na/Ca exchange, *Am. J. Physiol.* **236**(3), C103–110.

Mullins, L. J., and Brinley, F. J., Jr. (1975). Sensitivity of calcium efflux from squid axons to changes in membrane potential, *J. Gen. Physiol.* **65**, 135–152.

Mullins, L. J., and Requena, J. (1981). The "late" Ca channel, *J. Gen. Physiol.* (in press).

Requena, J., Mullins, L. J., and Brinley, F. J., Jr. (1979). Calcium content and net fluxes in squid giant axon, *J. Gen. Physiol.* **73**, 327–342.

Rojas, E., and Taylor, R. E. (1975). Simultaneous measurements of magnesium, calcium and sodium influxes in perfused squid giant axons under membrane potential control, *J. Physiol.* **252**, 1–27.

Shoukimas, J., Adelman, W. J., Jr., and Sege, V. (1977). Cation concentrations in the hemolymph of Loligo Pealei, *Biophys. J.* **18**, 231–234.

10

A Quantitative Expression of the Electrogenic Pump and Its Possible Role in the Excitation of *Chara* Internodes

UICHIRO KISHIMOTO, NOBUNORI KAMI-IKE, and
YŪKO TAKEUCHI

Chara and *Nitella* are freshwater algae. These two algae are very close to each other in taxonomic terms. The diameter of the internode cell of *Chara corallina* we have been using is about 0.6 mm and the length is about 6 cm on the average. We keep the internode cells after isolating from adjacent cells in artificial pond water, APW. APW is a solution having the following composition: 0.05 mM KCl, 0.2 mM NaCl, 0.1 mM Ca(NO$_3$)$_2$, and 0.1 mM Mg(NO$_3$)$_2$. The pH of APW was adjusted with 2 mM TRICINE (tris hydroxymethyl-methylglycine) or 2 mM MES(2-N-morpholionethane sulfonic acid).

The idea that an electrogenic pumping mechanism should exist in *Chara* and *Nitella* plasma membranes is based on the following observations: (1) The membrane potential is more negative than the calculated

UICHIRO KISHIMOTO, NOBUNORI KAMI-IKE, and YŪKO TAKEUCHI • Department of Biology, College of General Education, Osaka University, Osaka, Japan.

Nernst potential for the gradient across the plasmalemma for each of the following ions: K^+, Na^+, Cl^-, and OH^-. (2) The membrane potential is sensitive to light, temperature, and also to metabolic poisoning.

Kitasato showed in 1968 that the membrane potential of *Nitella* was sensitive to changes in external pH. However, it was about 70 to 80 mV more negative than that calculated with the Goldman equation at each pH. Moreover, the sum of the conductances for K^+, Na^+, and Cl^-, which was calculated from the data of Kitasato's flux measurements by using a constant field assumption, was negligibly small compared with the conductance measured electrically. He supposed that the excess hyperpolarization of the membrane potential was caused by an active pumping out mechanism of H^+ at the plasmalemma. Slayman (1965) had already proposed the idea of an H^+ pump from his experiments on *Neurospora hyphae* membranes. Recently he and his collaborators (1973) showed that the membrane potential of the *Neurospora* membrane was dependent on the internal ATP concentration.

The idea of an H^+ pump had been suggested previously for membranes of mitochondria and chloroplasts and also for the plasma membranes of some prokaryotes (Mitchell, 1963). The possible existence of an H^+ pump in the plasma membrane of higher plants was taken up by plant physiologists to explain the results of their experiments on ion transport and on the membrane potential (see Spanswick, Lucas, and Dainty, 1979). Shimmen and Tazawa (1977) showed in *Chara* that the membrane potential was depolarized to the level of the diffusion potential, i.e., -100 mV, when the internal ATP level was decreased down to 20 μM or less by using a technique of internal perfusion.

The data reported so far show a general similarity. However, there has been a diversity of conclusions drawn from these data. One reason for this diversity seems to come from workers using different definitions of the activity of the pump. Another reason arises from the differences in the way ionic conductances have been estimated. What seems essential for characterizing the electrogenic pump is to measure the ionic conductance of the *Chara* membrane as accurately as possible.

We applied a small test square wave current pulse under the condition of current clamp. That is, the current was kept zero except during application of the test current pulse. The current clamp technique was applied following Cole and Moore's method (1960). The arrangement of two pairs of electrodes, one for current monitoring another for

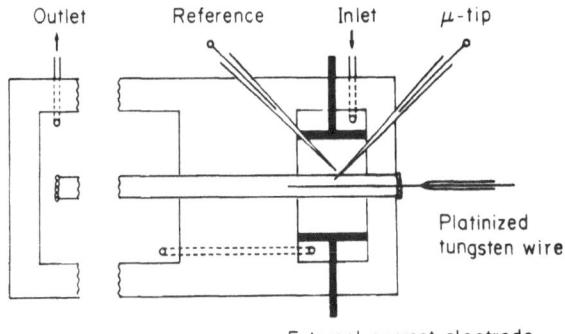

Figure 1. Electrode arrangement in the current clamp for *Chara* internodes. The internal axial wire is a platinized tungsten wire of 100 μm in diameter, the tip of which was sharpened electrolytically in advance. The external current electrode is a chlorinated silver plate, the surface of which was additionally plated with platinum black to give both stability of the electrode potential and low electrode impedance (Cole and Kishimoto, 1962).

voltage recording, is shown in Figure 1. The shape of the corresponding voltage response could be simulated satisfactorily with the models shown in Figure 2. We applied a least-squares method with the aid of a microcomputer for the determination of the series resistance, r_s, membrane resistance, r_m, and membrane capacitance, c_m. The additional parameters r and c, in the better simulation with a two time constant model (Figures 2b and 2c), are likely to arise from the contribution from the ionic processes of Cl^- and K^+ current flows. More details are described in our recent report (Kishimoto, Kami-ike, and Takeuchi, 1980).

We noticed in *Chara* that the two models illustrated, respectively, in Figures 2a and 2c gave the same value for r_m, if a short-duration (i.e., about 40 msec) and low-amplitude current, which causes voltage responses of less than 3 mV, was applied. By applying this method of stimulation for each voltage response we could determine both the change of the ionic conductance, G, which is the reciprocal of r_m in Figure 2c, and the change of the electromotive force, E, during the process of metabolic poisoning (Figure 4a). The data of the following inhibitor experiments are taken from our recent report (Kishimoto, Kami-ike, and Takeuchi, 1980)

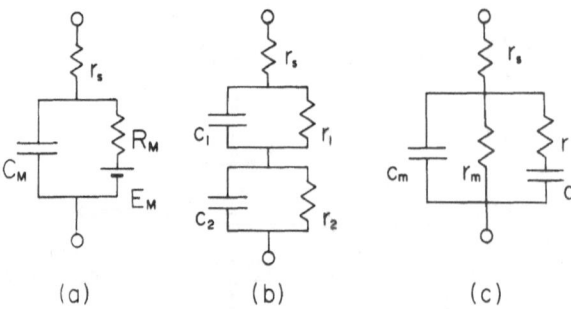

Figure 2. (a) General circuit model for the *Chara* membrane. The simulation of the voltage response with this model is generally satisfactory, if the amplitude of the voltage change under the current clamp is less than 3 mV. In some cases and especially if the voltage response is larger than 3 mV, better simulation could be achieved only with models having two time constants [(b) and (c)]. The simulation was first performed with a model shown in (b). Later, this model was transformed into another one shown in (c). The additional parameters r and c in (c) are highly likely to arise from the contribution of the ionic processes of Cl^- and K^+ current flows. Such changes cause actually an additional change of the electromotive force [E_m in (a)] of the membrane during the test current pulse. The reciprocal of r_m in (c) was adopted as the ionic conductance, G, of the *Chara* membrane.

CONDUCTANCES AND ELECTROMOTIVE FORCES DURING THE PROCESS OF INHIBITION OF THE ELECTROGENIC PUMP WITH 2 μM TRIPHENYLTIN CHLORIDE (TPC)

We carried out the inhibitor experiments in darkness in order to avoid any additional effects resulting from photosynthesis. The pH of the external solution was kept at 7. The temperature was kept at 25°C by using a thermoelectric transducer.

One of the general aspects of metabolic poisoning is a gradual decrease of G to a smaller nonzero level. Another aspect is a gradual shift of E to a less negative level. This effect may be seen in Figure 4a.

Triphenyltin chloride, TPC, is known to be a specific inhibitor of energy transduction at Fl particles in mitochondrial membranes and CFl particles in chloroplast membrane. TPC was dissolved in the APW, and the pH of the solution was buffered with 2 mM MES to 7.

A typical example* of the process of poisoning with 2 μM TPC is shown in Figures 4 and 5. As mentioned above, the decrease of G and the

*The data of the effect of 2 μM TPC (and its analysis) are taken from a report by U. Kishimoto, N. Kami-ike, and Y. Takeuchi (1980). The role of electrogenic pump in *Chara corallina, J. Membr. Biol.* **55**, 149.

Figure 3. A circuit model for the *Chara* membrane having the electrogenic ion pumping system beside the passive diffusion channel. The conductance, G, in Figure 2 is the sum of the conductances g_1 and g_2, which correspond to the conductances of the passive diffusion channel and of the electrogenic pump channel, respectively.

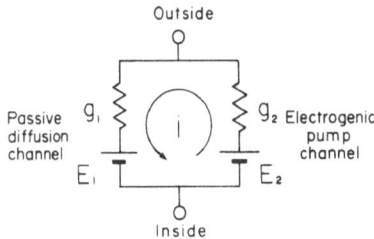

shift of E to a less negative level was reversible upon removing the inhibitor. Therefore, it is reasonable to suppose that the final asymtotic levels of G and E at the late stage of poisoning correspond to the conductance g_1 and the electromotive force E_1 of the passive channel.

The simplest and most reasonable model for this phenomenon is the one shown in Figure 3. Such a model was adopted in 1951 by Ussing and Zerahn for their estimation of active sodium flux in the short-circuited frog skin, and also in 1970 by Spanswick for the demonstration of the electrogenic pump in *Nitella*. This type of model for an ion pump having conductance has also some theoretical grounds (Finkelstein, 1964; Rapoport, 1970).

Under an implicit assumption that the passive channel is not influenced by metabolic poisoning the following equations can be written:

Before inhibition:

$$G(t)=g_1+g_2(t) \tag{1}$$

$$E(t)=[g_1E_1+g_2(t)\cdot E_2(t)]/G(t) \tag{2}$$

where G, E, g_1, and E_1 are defined above, and g_2 and E_2 are the conductance and the electromotive force of the pump channel, respectively.

At the late stage of inhibition:

$$G(t)=g_1 \tag{3}$$

$$E(t)=E_1 \tag{4}$$

Thus,

$$g_2(t) = G(t) - g_1 \tag{5}$$

$$E_2(t) = [G(t) \cdot E(t) - g_1 \cdot E_1]/g_2(t) \tag{6}$$

Pump current:

$$i(t) = \frac{g_1 \cdot g_2(t)}{g_1 + g_2(t)} \cdot [E_2(t) - E_1] \tag{7}$$

The amount, $i(t)/g_1$, corresponds to the extent of excess hyperpolarization, which is caused by the electrogenic pump. This amount has been regarded as an index of the pump activity.

Changes of g_2 and E_2 during TPC poisoning were calculated with these equations and the results are shown in Figure 4. The results are summarized as follows: (1) an exponential decay of g_2 to almost zero, (2) a shift of E_2 to a less negative level after the transient hyperpolarization, and (3) an exponential decrease of the pump current.

Generally, the action potential is supposed to be an event which occurs at the passive channels. Therefore, it may not be unreasonable to assume that the process of excitation does not affect the pump mechanism. Since we already know $g_2(t)$ and $E_2(t)$ of the pump channel during

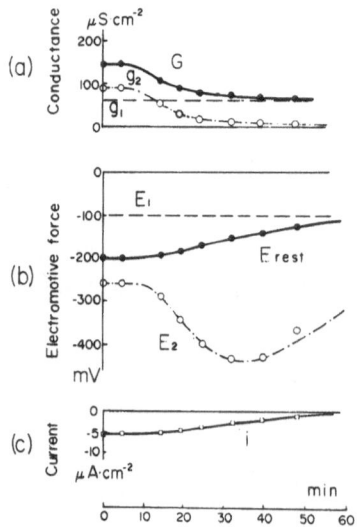

Figure 4. Changes of conductances (a), electromotive forces (b), and pump current (c) of the *Chara* membrane during the process of TPC (2 μM) poisoning. Temperature was 25°C and pH of the APW was adjusted to 7 with 2 mM MES buffer. The conductance, g_1, and the electromotive force, E_1, of the passive diffusion channel are decided as the final asymptotic levels of the measured conductance, G, and the electromotive force, E, during TPC poisoning. The conductance g_2 and the electromotive force, E_2, of the electrogenic channel and the pump current, i, were calculated with equations (5), (6), and (7) in the text.

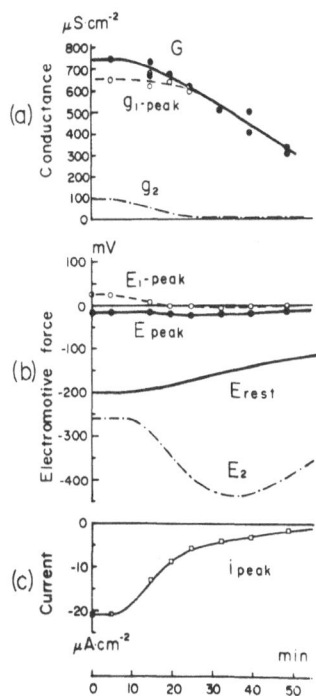

Figure 5. Changes of conductances (a), electromotive forces (b), and pump current (c) at the peak of action potential of *Chara* membrane during the process of TPC ($2 \mu M$) poisoning. The changes of g_2 and E_2 of the electrogenic pump channel are assumed to be the same as those in Figure 4. A large value of G at the peak of action potential is mainly due to g_1 of the passive diffusion channel. Note that E_1 at the peak of the action potential did not change appreciably by TPC poisoning. The pump current at the peak of the action potential was about 5 times as large as that at rest before TPC poisoning and decreased to almost zero with the progress of TPC poisoning.

TPC poisoning, it is possible to calculate g_1 and E_1 of the passive channel during the action potential.

The changes of g_1 and E_1 at the peak of the action potential during the process of poisoning were calculated with equations (1) and (2), and these are shown in Figure 5. The results can be summarized as follows: (1) The main part of the marked increase of G at the peak of the action potential is due to g_1. (2) E_1 at the peak of the action potential is almost not influenced by TPC. (3) E_1 shows an overshoot before and during the early stage of TPC poisoning. (4) At the peak of the action potential the pump current is 4–5 times as large as that at rest at the beginning of poisoning and decayed to almost zero in the later phase of TPC poisoning.

The large value of g_1 at the peak of the action potential decreased slowly with the progress of poisoning. If this were not the case, TPC would be an ideal inhibitor for the pumping mechanism in *Chara* membranes. Nevertheless, TPC is much better than other inhibitors so far tested (such as $0.2 \, mM$ DNP, $5\mu M$ CCCP, $5 \, \mu M$ DCMU, etc.). These other inhibitors caused not only a reduction of g_1 at the peak of action potential, but also a shift of E_1 at the peak to a more negative level which was close to E_1 at rest. In other words, the *Chara* membrane became inexcitable at the late stage of poisoning with these inhibitors. On the other hand, with 2 μM TPC the action potential could be elicited by stimulation even during the late stage of inhibition.

Changes of g_1, E_1, and i during an action potential are compared before and at the late stage of TPC poisoning in Figures 6a and 6b. The

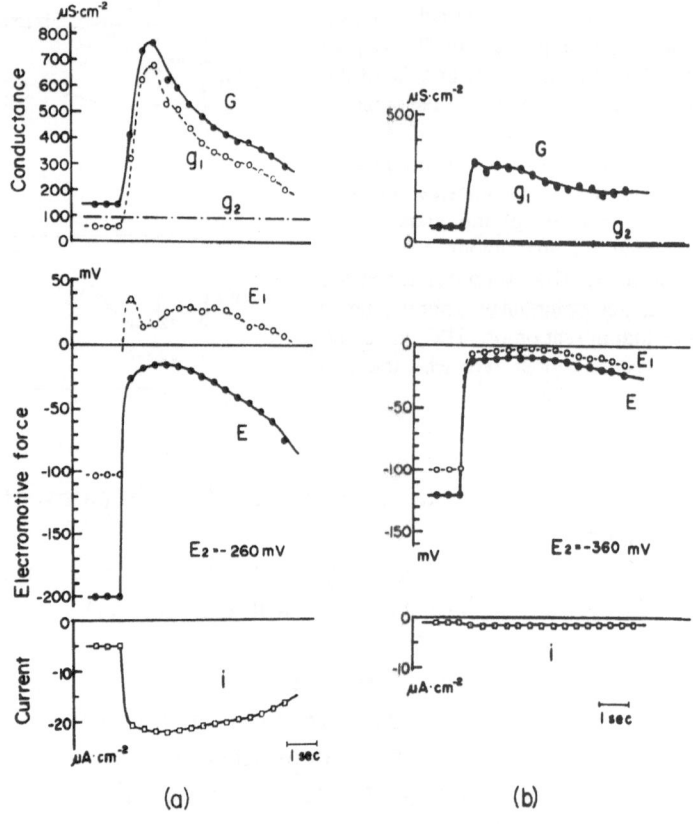

Figure 6. Changes of conductance, electromotive force, and pump current during action potential of *Chara* membrane before (a) and at the late stage of TPC poisoning (b). Temperature was 25°C and pH was adjusted to 7 with 2 mM MES buffer. The main part of the conductance changes is due to g_1 of the passive diffusion channel. The change of E is almost in parallel with that of E_1 of the passive diffusion channel. The values of E_2 in (a) and (b) were taken from Figure 5 in these two cases and were assumed to be unchanged during action potential. The pump current, i, increased rapidly to a large maximum at the peak of the action potential and began to decrease slowly at around its falling phase. Note that the pump current decreased markedly and the duration of the action potential was greatly prolonged at the late stage of TPC poisoning.

interesting feature to note is that the duration of the action potential was prolonged markedly at the late stage of TPC poisoning. This may suggest that the large pump current, i.e., about 20 $\mu A cm^{-2}$, at the peak of the action potential before poisoning flows back into the passive excition channels, causing an inactivation of the excitatory process there. On the other hand, it cannot do so at the late stage, since the pump current decreased to almost zero. However, we did not observe such a prolongation of the action potential with 0.2 mM DNP poisoning. In this case, the *Chara* membrane was inexcitable at the late stage of inhibition. A similar prolongation of the action potential was observed, when the temperature was decreased from 25°C down to 5°C (unpublished observations). About a 40% decrease of G, a 20% depolarization of E, and a marked decrease of the pump current were also found at 5°C. However, such an prolongation of the action potential at low temperature could also be a direct temperature effect on the excitatory mechanism and might not be caused by a decrease in the pump current.

It is also worth mentioning in the TPC experiments that G and E recovered in 10–20 min even without removing TPC from the external solution, if we illuminated the specimen (i.e., 2000 lux by an incandescent lamp).

INTERNAL ATP LEVEL DURING TPC POISONING

The internal ATP level of the *Chara* cell was also measured by using a Luciferin–Luciferase reaction. The ATP level decreased with the progress of TPC poisoning (Figure 7). However, it never became zero, but remained at 200–300 μM, which was 1/3 to 1/4 of the original level. This indicated that the electrogenic pumping activity was almost completely depressed, even though the internal ATP remained at 200–300 μM. A similar result was found by Keifer and Spanswick (1979).

THE pH DEPENDENCE OF CONDUCTANCES AND ELECTROMOTIVE FORCES

What factors are determining g_2 and E_2 of the pump channel, and what is the reason for the transient hyperpolarization of E_2 during metabolic poisoning? Since the electrogenic pump is likely to be an ATP-driven H^+ pump, it seems straightforward to check the pH dependence of g_2 and E_2. So far, Kitasato (1968), Richard and Hope (1974), Keifer and Spanswick (1978), and Fujii, Shimmen, and Tazawa (1979) have reported on the pH dependence of the membrane potential of *Nitella* and *Chara* cells making use of ionic conductance data. These results have much in common. However, there are also several subtle differences among these data. Such differences seem to come partly from differences in the way of determining conductances and partly from differences in models adopted for the electrogenic pump.

We have measured G and E as functions of the external pH before and after TPC treatment. One example is shown in Figure 8. As shown in Figure 8a, g_1 was almost unchanged with the increase of external pH from 5.4 to 8.3. Beyond pH 8.3 it increased. E_1 moved to more negative levels with the increase of pH up to 8.3. The maximum slope was 27 mV/pH. Beyond pH 8.3, E_1 was depolarized again. On the other hand, g_2 showed almost a linear decrease with pH from 6.3 to 8.3. The g_2 increased again when the pH was raised above 8.3. E_2 showed almost a linear shift to more negative levels when the pH was increased from 5.4 to 8.3. The slope of the E_2 change was about 52 mV/pH unit. Beyond pH 8.3 E_2 was depolarized again.

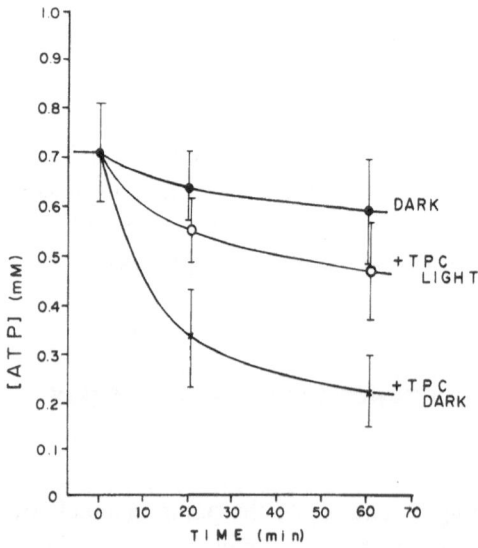

Figure 7. Changes of internal ATP concentration in the dark with (×) and without (●) 2 μM TPC and in the light (○) with 2 μM TPC. Temperature was 25°C and pH was adjusted to 8 with 2 mM TRICINE buffer.

Figure 8. Changes of conductances (a), electromotive forces (b), and pump current (c) of *Chara* membrane with external pH change. The pH of the external APW was adjusted with 2 mM TRICINE or 2 mM MES. The data are on a single *Chara* internode. That is, the first series of measurement at different pH solution was carried out without 2 μM TPC in the light. Then, the second series of measurement was carried out on the same cell after poisoning with 2 μM TPC in the dark. The values of g_1 and E_1 were taken from those in the second series of measurements. From these two series of measurements g_2 and E_2 of the electrogenic pump channel can be calculated with equations (5), (6), and (7). See text for details.

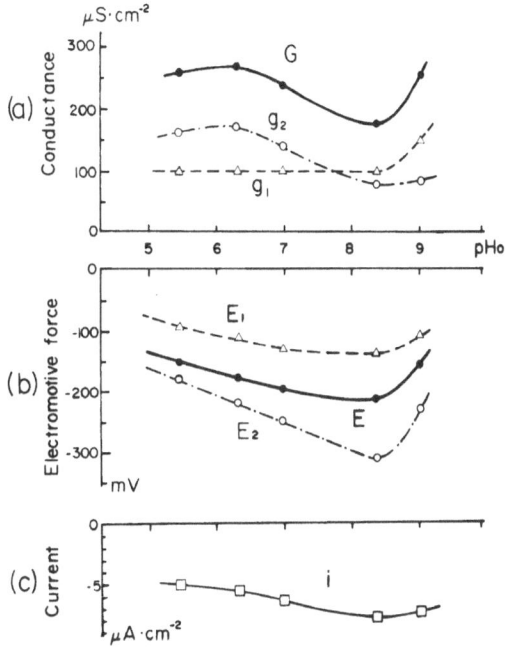

A MODEL FOR THE MECHANISM OF ELECTROGENIC PUMP

Here we make an assumption that the electrogenic pump of the *Chara* membrane is a result of coupled reactions of ATP hydrolysis and H^+ extrusion or uptake. Both are chemical reactions accompanied by respective free energy changes. Both reactions could be accompanied by charge movements (either H^+, OH^-, or electron) across the membrane. This situation can be modeled with the equivalent circuit shown in Figure 12a. This is tantamount to supposing that the electrogenic H^+ pumping-out reaction is a linear combination of two charge-carrying processes. That is,

$$E_2 = \left(\frac{g_3}{g_2} \right) \frac{RT}{F} \left[\ln K_{\text{ATP}} - \ln \frac{\text{ATP}_i}{\text{ADP}_i \cdot P_i} \right]$$

$$+ \left(\frac{g_4}{g_2} \right) \frac{RT}{F} \left[\ln K_{\text{H}} - (\text{pH}_o - \text{pH}_i) \right] \tag{8}$$

$$g_2 = g_3 + g_4 \tag{9}$$

where K_{ATP} ($=6.7\ \mu M$) is the dissociation constant of ATP. ATP_i, ADP_i, and P_i are the internal concentrations of ATP, ADP, and P, respectively. The external and internal pH are shown as pH_o and pH_i, respectively. Conductances of these two channels are expressed as g_3 and g_4. These conductances can also be regarded as the extent of the contribution of each reaction to the electrogenic pump. The value of $RT\ \ln K_H$ in the second term in equation (8) is the standard free energy change of channel 4, which can be determined experimentally.

The ratio g_4/g_2 can be determined from the slope of the change in E_2 plotted against external pH (Figure 9b). The value of g_3 can be calculated, since $g_2 = g_3 + g_4$. E_3 is estimated to be -525 mV and is assumed to be unchanged with changes in external pH. Then, the value

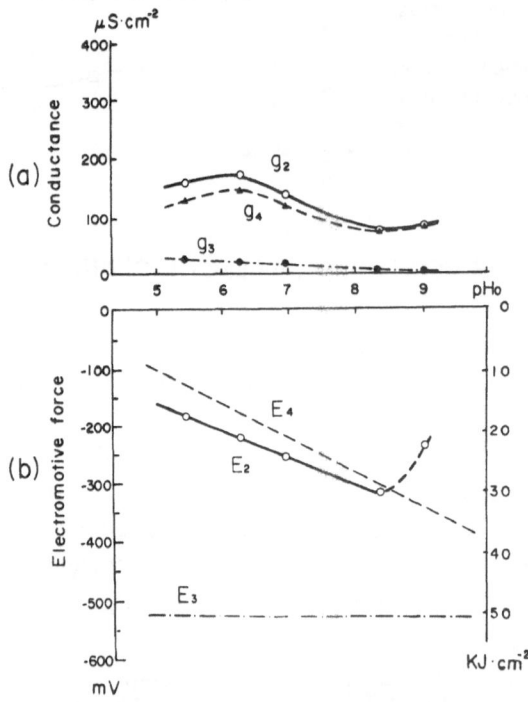

Figure 9. Conductance (a) and electromotive force (b) of the electrogenic pump channel was further divided into two channels, one being driven by ATP hydrolysis and another dependent of the pH difference across the *Chara* membrane. The situation is expressed with equations (9) and (10) in the text and in Figure 12a. The conductances g_3, g_4 and electromotive forces E_3 and E_4 can be determined by knowing g_2 and E_2 and the slope of E_2 for instance at pH 7. See text for details.

of $(RT/F)\ln K_H$ can be decided. This value is -216 mV, if E_2 at pH 7 is -256 mV (see Figure 9b). For some unknown reason, we have found that this value has been different among samples so far tested.

Anyway, after determining E_4 at pH 7, E_4 in other pH solution can be calculated with equation (10)

$$E_4 = -216 - 59(\text{pH}_o - \text{pH}_i) \quad (\text{mV}) \tag{10}$$

Once $E_3(= -525 \text{ mV})$ and E_4 have been decided in this way, g_3 and g_4 at each pH can be calculated. The results are shown in Figure 9b.

The level of E_2 of the electrogenic pump should be between E_3 and E_4 at each pH, if E_3 remains constant and E_4 changes following equation (10) as is shown in Figure 9b. However, E_2 was depolarized beyond the expected level of E_4 at pH 9. This was caused by the depolarization of E_4 from that expected from equation (10) and also by the depolarizations of E_3 as will be discussed later. Therefore, it is evident that some selectivity change should have occurred in the membrane upon exposure to solutions whose pH values were larger than 8.3.

Once we have determined values of E_3 and E_4 at a known pH, $g_3(t)$ and $g_4(t)$ at each time during the process of TPC poisoning can be calculated with a procedure similar to that described in the preceding paragraph. In the TPC experiment, both external and internal pH were almost equal to 7. Thus, adopting -525 mV for E_3 (as described before) and -120 mV for E_4 in this internode cell, $g_3(t)$ and $g_4(t)$ were calculated and these are shown in Figure 10. Practically, $g_3(t)$ and $g_4(t)$ were simulated as exponential processes. It is evident that $g_3(t)$ decays with a comparatively slow rate to zero during the process of poisoning. On the other hand, $g_4(t)$ decays rapidly to a small nonzero level. The reason for the transient hyperpolarization of E_2 during metabolic poisoning now can be explained. That is, $g_3(t)$ can be larger than $g_4(t)$ in a certain period of metabolic poisoning. This period certainly corresponds to the period of the transient hyperpolarization of E_2.

THE pH DEPENDENCE OF PUMP CURRENT

Coming back again to the passive channel, the changes of the electromotive force E_1 against external pH change could be simulated

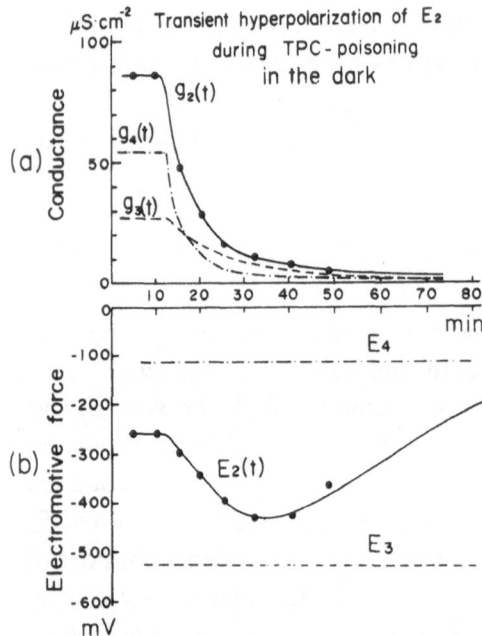

(a)

(b)

Figure 10. Transient hyperpolarization of the electromotive force, E_2, of the electrogenic pump channel during TPC poisoning. Changes of g_3 and g_4 of the electrogenic pump channel were calculated by knowing g_2 and E_2 at each time during TPC poisoning (a). Here, E_3 was assumed unchanged during poisoning for simplicity and change of E_4 was calculated with the second term of equation (8). The value of g_3 was larger than that of g_4 in a certain period of poisoning, which caused a transient hyperpolarization of E_2 (b).

Figure 11. Selectivity change of the passive diffusion channel of the *Chara* membrane for K^+ and Cl^- and/or H^+ against external pH change. The largest symbols ($\bigcirc$, $\triangle$, and $\square$) are conductances for K^+, Cl^-, and H^+, respectively, decided by simulation of changes of E_1 and g_1 under an assumption of no H^+ flow through the passive diffusion channel. The smallest symbols are conductances determined similarly under an assumption that all the pump H^+ current flowed back into the cell through the passive diffusion H^+ channel. The middle-sized symbols are conductances determined similarly under an assumption that 4/5 of the pump H^+ current flowed back through the passive H^+ channel. Actual situation may be between the two extreme cases. Anyway, the selectivity of the *Chara* membrane tended to be lost at pH above 8.3.

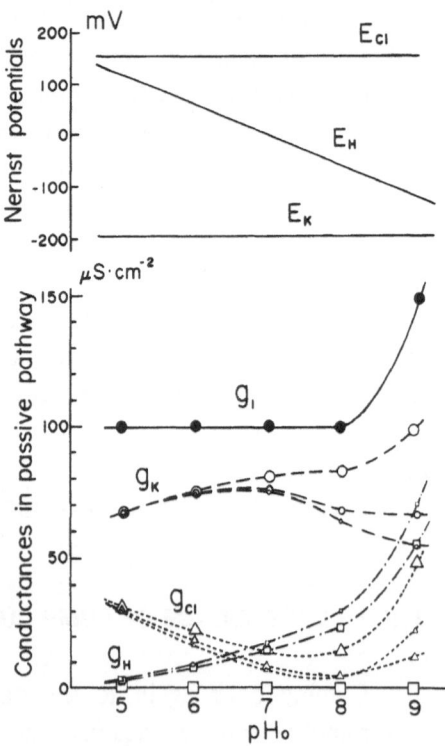

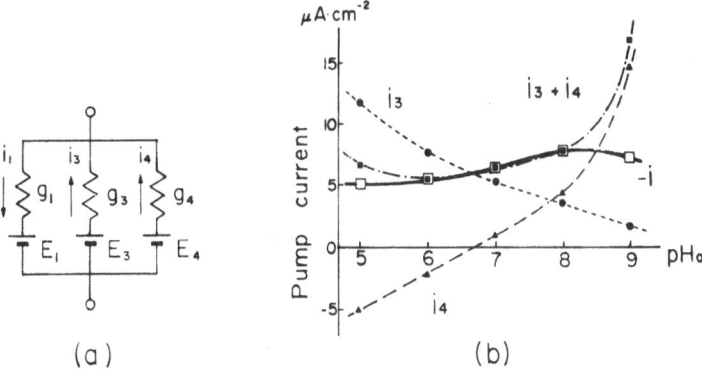

Figure 12. The pH dependence of the current through the electrogenic pump channel(s) and the passive diffusion channel. The values of g_3, g_4, E_3, and E_4 were determined as described in Figure 9. The sum of the pump currents ($i_3 + i_4$) is equal to i in Figure 8 for the external pH between 6 and 8.3. However, the sum ($i_3 + i_4$) is much larger than i in Figure 8 at pH above 8.3. The reason comes from the assumption of constant value of $\ln k_H$ in equation (8) in the calculation of E_4. A conclusion we draw from this pH experiment is that E_4 depolarized at pH above 8.3. This suggests that a qualitative change occurred in the *Chara* membrane at such a high-pH solution.

with the Goldman equation. It was necessary to assume a very high permeability constant for H^+ and its probable change with pH. A similar simulation was also successful using an electrical circuit model, in which g_K and E_K, g_{Cl} and E_{Cl}, and/or g_H and E_H were assumed to exist in parallel. These two simulations showed that the permeabilities or conductances for K^+ and Cl^- greatly increased above pH 8.3 (Figure 11). We frequently observed a spontaneous action potential, when we changed the pH of the external solution from 7 to 9 or 10.

The situation might become more clear if we express the ion conducting channels, i.e., passive and active, in the manner shown in Figure 12a. The currents which flow through these three channels can be calculated from the data in Figures 8 and 9, and these are shown in Figure 12. Generally satisfactory agreement between the expected (i.e., $i_3 + i_4$) and the calculated [i.e., $i = g_1(E_2 - E_1)$] values were found between pH 6 and pH 8.3. The large discrepancy between these two values at pH 9 comes from the estimation of E_4 with equation (10). If we suppose that E_4 was depolarized to -227 mV at pH 9 rather than to the value expected (i.e., -333 mV) from equation (10), such a discrepancy disappears, as mentioned earlier. We can draw out the same conclusion that E_4 should have been depolarized from the assumed -333 mV to less than -233 mV at pH 9 in order that g_3 remain as a positive value. Therefore, it is highly likely that the pump channel, as well as the passive channel, changes its selectivity at pH values above 8.3. Recovery from this change by lowering external pH takes many hours.

So far we have supposed that the electrogenic pump in the *Chara* membrane is an ATP-driven H^+ pump. However, it could be an OH^- pump as well. It is also possible that both types of pumps are operative. We also are aware that Cl^-, Na^+, HCO_3^-, Ca^{++}, etc. are actively transported in *Chara* membranes. The relation of the active transport of these ions to the H^+ pump is to be investigated quantitatively in the near future. The complex but important role of photosynthesis in *Chara* has not been discussed in this report.

ACKNOWLEDGMENT

This work was supported by a Research Grant from the Ministry of Education, Science and Culture of Japan.

REFERENCES

Cole, K. S., and Kishimoto, U. (1962). Platinized silver chloride electrode, *Science* **136**, 381.

Cole, K. S., and Moore, J. W. (1960). Ionic current measurements in the squid giant axon membrane, *J. Gen. Physiol.* **44**, 123.

Finkelstein, A. (1964). Carrier model for active transport of ions across a mosaic membrane, *Biophys. J.* **4**, 421.

Fujii, S., Shimmen, T., and Tazawa, M. (1979). Effect of intracellular pH on the light-induced potential change and electrogenic activity in tonoplast-free cells of *Chara australis*. *Plant and Cell Physiol.* **20**, 1315.

Keifer, D. W., and Spanswick, R. M. (1978). Activity of the electrogenic pump in *Chara corallina* as inferred from measurements of the membrane potential, conductance and potassium permeability, *Plant Physiol.* **62**, 653.

Keifer, D. W., and Spanswick, R. M. (1979). Correlation of adenosine triphosphate levels in *Chara corallina* with the activity of the electrogenic pump, *Plant Physiol.* **64**, 165.

Kishimoto, U., Kami-ike, N., and Takeuchi, Y. (1980). The role of electrogenic pump in *Chara corallina*, *J. Membr. Biol.* **55**, 149.

Kitasato, H. (1968). The influence of H^+ on the membrane potential and ion fluxes of *Nitella*, *J. Gen. Physiol.* **52**, 60.

Mitchell, P. (1963). *The Structure and Function of the Membrane and Surfaces of Cells*, D. J. Bell and J. K. Grant, Ed. (Cambridge Univ. Press, London), pp. 142–169.

Rapoport, S. I. (1970). The sodium–potassium exchange pump: Relation of metabolism to electrical properties of the cell. I. Theory, *Biophys. J.* **10**, 246.

Richard, J. L., and Hope, A. B. (1974). The role of protons in determining membrane electrical characteristics in *Chara corallina*, *J. Membr. Biol.* **16**, 121.

Shimmen, T., and Tazawa, M. (1977). Control of membrane potential and excitability of *Chara* cells with ATP and Mg^{2+}, *J. Membr. Biol.* **37**, 167.

Slayman, C. L. (1965). Electrical properties of *Neurospora crassa*: respiration and the intracellular potential. *J. Gen. Physiol.* **49**, 93.

Slayman, C. L., Long, W. S., and Lu, C. Y.-H. (1973). The relationship between ATP and an electrogenic pump in the plasma membranes of *Neurospora crassa*, *J. Membr. Biol.* **14**, 305.

Spanswick, R. M. (1970). The effects of bicarbonate ions and external pH on the membrane potential and resistance of *Nitella translucens*, *J. Membr. Biol.* **2**, 59.

Spanswick, R. M., Lucas, W. J., and Dainty, J. (1979). *Plant Membrane Transport: Current Conceptual Issues* (Elsevier/North-Holland, Amsterdam).

Ussing, H. H., and Zerahn, K. (1951). Active transport of sodium as the source of electrical current in the short-circuited isolated frog skin. *Acta Physiol. Scand.* **23**, 110.

Part IV

Stimuli and Drugs

11

Increases in Membrane Conductance Caused by Electrical, Chemical, and Mechanical Stimuli

GERALD EHRENSTEIN and GIDEON GANOT

INTRODUCTION

We would like to consider two basic questions regarding the mechanisms by which electrical, chemical, and mechanical stimuli affect the nervous system. These questions are as follows:

(1) Which membrane circuit element is changed by these stimuli?
(2) What type of molecular change underlies the circuit element change?

The first question has already been clearly answered, and the purpose of considering it now is to compare the answer with the

GERALD EHRENSTEIN • Laboratory of Biophysics, National Institute of Neurological and Communicative Disorders and Stroke, National Institutes of Health, Bethesda, Maryland 20205.
GIDEON GANOT • Department of Physiology, Technion-Faculty of Medicine, P.O.B. 9649, Haifa, Israel.

possibilities. The second question has been answered for electrical and chemical stimuli, and one purpose of this presentation is to clarify the possibilities with regard to mechanical stimuli.

In retrospectively comparing possibilities with known results, I am reminded of the three stages of a good, new idea. First, it is passed off as ridiculous. Secondly, it is allowed that it may be possible, but it really is rather far-fetched. Finally, it is agreed that although it is correct, that is not very surprising since there really was not any reasonable alternative in the first place.

THE ROLE OF MEMBRANE CONDUCTANCE CHANGES

It is now well established that it is membrane conductance that changes in response to an electrical stimulus. Kacy Cole, whose eightieth birthday we are celebrating, had a central role in this discovery. Cole and Curtis (1939) showed more than 40 years ago that during an action potential there is an increase in membrane conductance, as evidenced by the balancing and unbalancing of an electrical bridge. As shown in Figure 1, the logo for *The Annual Review of Biophysics and Bioengineering* is based on this elegant demonstration. Kacy was also the central figure in the invention of the voltage clamp (Cole, 1949), which has been so

Figure 1. Photograph of a portion of the cloth binding of Volume 9 of *Ann. Rev. Biophys. Bioeng.* showing the logo referred to in the text.

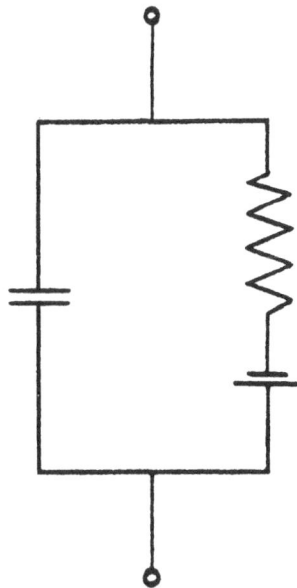

Figure 2. Simple equivalent circuit for a membrane between two nonidentical solutions.

successful in developing the precise description of the voltage-dependent conductance changes underlying electrical excitation.

In retrospect, and with present understanding of membrane structure, what properties of the system could be affected by an electrical signal? The simplest schematic diagram for an axonal membrane is shown in Figure 2. The parameters that could possibly change in response to an electrical stimulus are

1. driving force (represented by the battery in Figure 2)
2. membrane capacitance
3. membrane conductance

The driving force depends upon the ionic concentrations, and the ionic concentrations change very slowly. In fact, a giant axon can support many thousands of action potentials before the internal ionic concentration is changed significantly. Thus, driving force changes could not trigger relatively fast action potentials.

The specific capacitance of a membrane depends upon its composition and its thickness. For an axonal membrane, it is not possible to change the composition of a substantial fraction of the membrane in a short time. Furthermore, for a membrane of fixed composition, the thickness can change only a small amount because of the limited flexibility of the lipid bilayer.

Thus, retrospectively, arguments can be given that there is no reasonable alternative to the membrane conductance as the parameter that responds to electrical stimuli.

The situation with regard to chemical and mechanical stimuli is very similar. It has been demonstrated experimentally that membrane conductance is the parameter that changes significantly in response to chemical stimuli, as, for example, acetylcholine stimulation of muscle endplates (Tekeuchi and Tekeuchi, 1960). Membrane conductance has also been found to change significantly in response to mechanical stimulation as, for example, in lobster axons (Julian and Goldman, 1962) and in paramecium (Naitoh and Eckert, 1973). Also the same arguments described above in connection with electrical stimulation can be used retrospectively to rule out alternatives to membrane conductance increases for chemical and mechanical stimulation.

IONIC CHANNELS IN MEMBRANES

A number of molecular mechanisms have been invoked to explain the changes in membrane conductance. For the voltage-dependent case, these have included

1. Voltage-dependent phase transitions of the lipid bilayer
2. Carrier transport
3. Voltage-dependent channels

The first direct observation of discrete membrane conductance changes was made on lipid bilayers doped with EIM—a bacterial extract known to cause voltage-dependent conductance (Mueller and Rudin, 1963). Bean *et al.* (1969) added small amounts of EIM to the solution bathing a lipid bilayer. When the membrane was voltage clamped to -10 mV, they observed that the current increased as a function of time in a discrete manner (Figure 3). These discrete current jumps correspond to conductance changes of about 400 pS—a value so large as to require the interpretation that EIM acts as a channel, rather than a carrier.

Figure 3. Long-lived conductance steps during membrane interaction with EIM. Membrane composition, brain phospholipid, 1.5%; tocopherol, 15%. Polarizing potential, −15 mV. Temperature, 36°C. Chart speed, 3 in./min. EIM, to about 20×10^{-9} g/ml, was added to the reference compartment at the two points indicated. The recording was made on an X–Y recorder, necessitating the overlapping, sequential scans. From Bean et al. (1969).

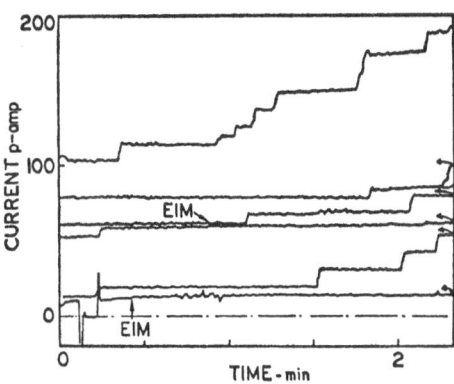

In terms of understanding the molecular mechanism underlying this observation, a key question is how the EIM channels respond to changes in membrane potential. The two extreme possibilities are as follows:

a. Each channel has two conductance states and switches between them in a manner depending on voltage.

b. Each channel has very many conductance states, and gradually changes conductance as membrane potential changes.

In the latter case, if the membrane potential were changed significantly, the conductance of a single channel would have essentially the same time course as a many-channel membrane. In the former case, however, the conductance of a single channel would change discretely between two levels. Either result would be consistent with the discrete conductance increases that had been found during addition of channels to the bilayer. The question was resolved by experiments on single EIM channels, which showed discrete changes between two levels, as shown in Figure 4 (Ehrenstein et al., 1970). Subsequent experiments on a number of other voltage-dependent channels in lipid bilayers have also shown discreteness in the conductance levels, although some channels have more than two states.

As summarized by Harold Lecar in Chapter 6 of this volume, single-channel experiments have now been performed on a number of biological membranes that respond to chemical stimulation. Recently,

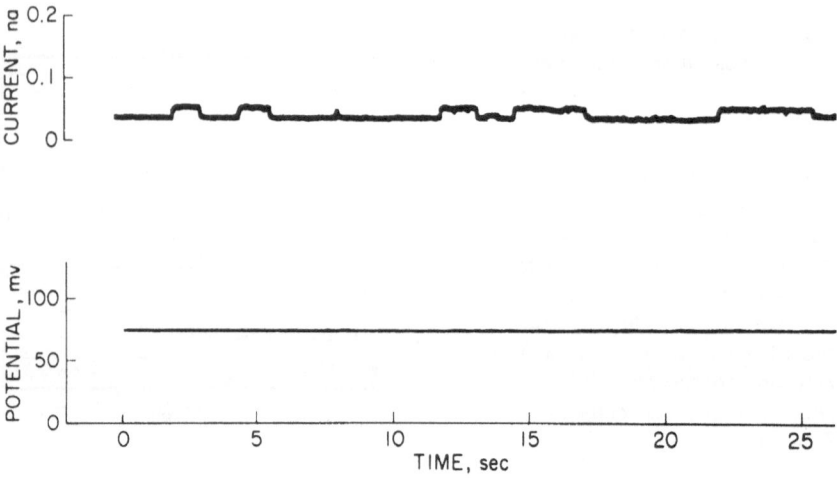

Figure 4. Discrete conductance jumps for a single EIM channel under voltage clamp. From Ehrenstein *et al.* (1970).

these observations were extended to a channel that can be stimulated electrically— the potassium channel of the squid, whose discrete conductance changes are shown in Figure 5 (Conti and Neher, 1980). The experimental results on both lipid bilayers and biological membranes clearly demonstrate that the primary mechanism for conductance change in membranes that respond to chemical and electrical stimulation is the discrete change in the conformation of specific membrane channels.

MECHANICALLY STIMULATED CHANGES IN MEMBRANE CONDUCTANCE

For mechanically-stimulated changes in membrane conductance, channels with discrete conformational changes have not yet been demonstrated by single-channel experiments or by measurements of noise power spectra. However, there are other kinds of evidence favoring discrete channels— at least for some types of mechanical stimulation. For example, the posterior mechanically stimulated response in paramecium can be blocked by tetraethylammonium (Naitoh and Eckert, 1973).

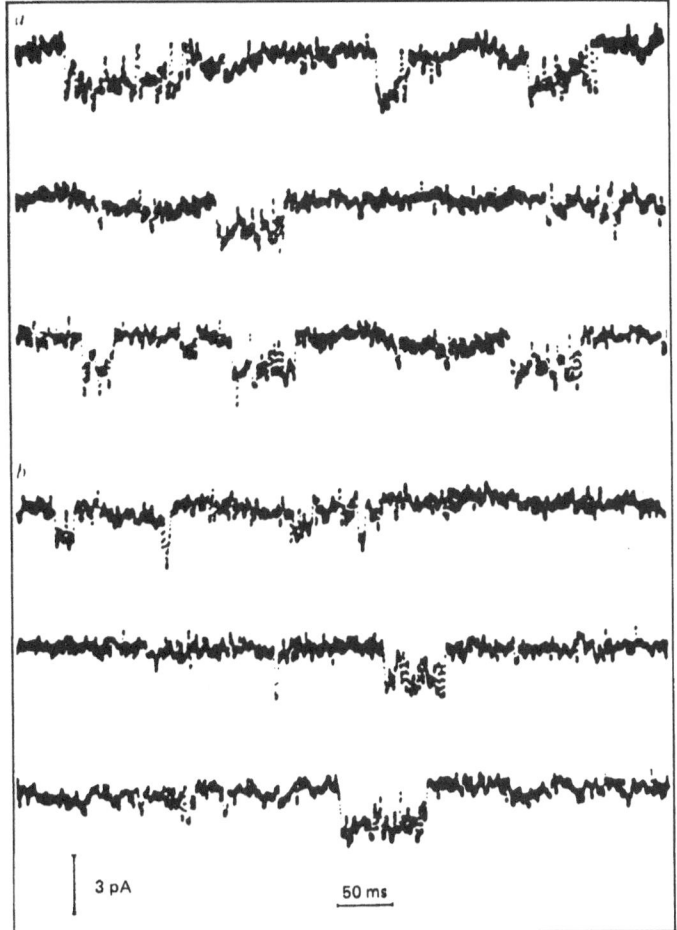

Figure 5. Discrete conductance jumps for single K$^+$ channels under voltage clamp. Under three traces were recorded at -25 mV membrane potential. Lower three traces were recorded at -35 mV. From Conti and Neher (1980).

Next, we will present evidence, from experiments on mechanical stimulation of axons, that an alternative mechanism is possible: mechanical stimulation can cause a gradual increase in the average size of nonspecific ionic pathways through the membrane. The advantage of using an axon preparation is that a wide range of electrical measurements have been developed for axons, and it has proved possible to combine some of them with mechanical stimulation. The disadvantage of using this preparation is that the axon does not normally respond to mechanical stimuli. Nevertheless, the axon might suggest mechanisms that could be employed physiologically.

Julian and Goldman (1962) showed that mechanical stimulation of the lobster axon caused a reversible depolarization (sometimes large enough to initiate an action potential) and an increase in membrane conductance. In order to demonstrate that the increase in membrane conductance is a direct result of the mechanical stimulus, they minimized the potential change during the response. As shown in Figure 6, the increase in membrane conductance is neither produced nor strongly influenced by changes in membrane potential.

We have followed up this work using *Myxicola* axons, and we would like to describe some of our preliminary results. A more complete account can be found in the paper by Ganot *et al.* (1981). When we applied a transverse mechanical stimulus by means of a plastic stylus driven by a loudspeaker, the axon became depolarized for several minutes. If the depolarization was large enough, it initiated an action potential.

In order to test whether the sodium channels responsible for action potentials are also involved in producing the generator potential for mechanical transduction, we added tetrodotoxin to the external solution. Tetrodotoxin is known to block these sodium channels and, as expected,

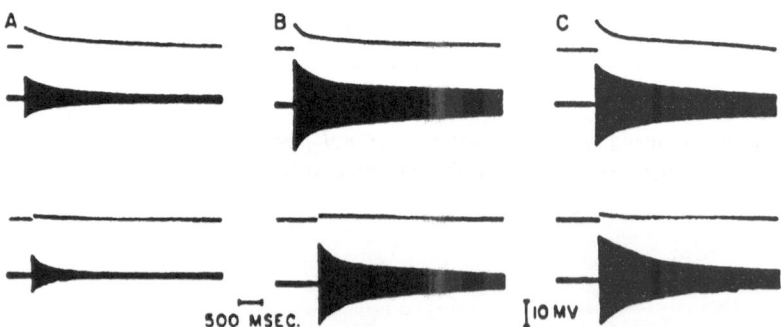

Figure 6. Effect of minimizing change on decrease in membrane resistance produced by mechanical stimulus. Parts A, B, and C show results from three different experiments. In each part, the upper records show magnitude of impedance change when no attempt was made to control potential. Lower records taken immediately after upper, but with current feedback used to minimize potential change produced by short mechanical stimuli of constant strength. Bridge output records retouched for greater clarity. From Julian and Goldman (1962).

it blocked the action potential. However, the mechanically induced depolarization was unaffected, indicating that these sodium channels are not involved in the mechanical transduction. Similar results have been reported for the pacinian corpuscle and the crayfish stretch receptor (Loewenstein *et al.*, 1963).

We also applied transverse mechanical stimulation to a voltage-clamped axon. In this system, the cross-sectional area of the stylus delivering the mechanical stimulus is much smaller than the area of the membrane under voltage clamp. When the stylus is activated, therefore, the change in conductance is not homogeneous over the voltage-clamped area. In general, this distorts the membrane current so that the measured conductance change is only approximately proportional to the actual conductance change. Only for the case where the mechanically induced current is zero does the system behave homogeneously. For this important case, the voltage clamp measurements are accurate. Therefore, accurate measurements of the reversal potential—the potential at which the mechanically induced membrane current changes sign—can be made.

The current induced by a mechanical stimulus applied when the axon is clamped to its resting potential increases to a peak value and then decays with a single exponential, suggesting there is a single process involved. In order to obtain the reversal potential for this process, we obtained current records for various clamped potentials with a mechanical stimulus of constant amplitude. In general, there was a steady-state current under voltage clamp before the mechanical stimulation was applied, and the mechanical stimulation then induced an additional component of current. The potential at which the peak amplitude of this mechanically induced current is zero is the reversal potential. For a mechanical stimulus of relatively small amplitude (29 μm) the reversal potential was -46 mV, and for a mechanical stimulus of relatively large amplitude (53 μm), the reversal potential was -3 mV. Additional experiments on deteriorated axons indicated that axonal damage caused the reversal potential to move in the hyperpolarizing direction. Thus, the change in reversal potential in the depolarizing direction caused by increasing the mechanical stimulus could not have been caused by axon deterioration.

The change in the reversal potential of the mechanically stimulated conductance corresponds to a change in selectivity. In principle, this could arise from a change in the relative number of excited channels of two or more different types—each with its unique reversal potential. This possibility is unlikely, however, since, as previously indicated, the conductance recovers with a single exponential. A more likely explanation is that the change in selectivity is related to a single conductance mechanism. Evidence indicates that biological channels are either open or closed, and hence have a fixed selectivity. In view of the observed decrease in selectivity with increasing amplitude of stimulus, we suggest that there is a different kind of ionic pathway, which gradually, rather than discretely, increases in diameter (and perhaps changes shape) in response to an increase in the amplitude of a mechanical stimulus.

The evidence we have presented that there are ionic pathways whose average diameter can be increased by mechanical stimulation raises the question as to whether these pathways would also transport ions in absence of mechanical stimulation. Such an ionic current could be all or part of what is called leakage current. Recent experiments in our laboratory have determined the reversal potential for leakage current in *Myxicola* axons by blocking the sodium and potassium channels with drugs (Ganot *et al.*, 1981). The leakage reversal potential was found to be about equal to the reversal potential reported here for a mechanical stimulus of small amplitude. This suggests that mechanical stimulation opens further a pathway that is already present—the leakage pathway.

A clue to the nature of the ionic pathways that can respond to mechanical stimulation is provided by a comparison of biological membranes with lipid bilayers. Leakage in biological membranes is several orders of magnitude larger than it is in lipid bilayers. Structurally, biological membranes differ from lipid bilayers in that they contain an abundance of membrane proteins. A reasonable explanation is that some of the membrane proteins provide the leakage pathway, perhaps by locally reorienting the lipids so that their polar ends swing into the membrane to provide a polar boundary for ions moving through the membrane. The average diameter of these pathways might increase when the membrane is stretched by mechanical stimulation. Although this may be considered a type of channel mechanism, it would allow relatively continuous changes in membrane conductance and selectivity.

Whether or not this speculation on the nature of the ionic pathways is correct, the evidence we have presented suggests that there are ionic pathways whose conductance changes in a gradual, rather than a discrete, manner in response to mechanical stimulation. It will be interesting to find out from future experiments whether this mechanism is utilized physiologically for mechanical transduction, or whether mechanical transduction, like electrical and chemical transduction, is always mediated by discrete conformational changes of channel proteins.

REFERENCES

Bean, R. C., Shepard, W. C., Chan, H., and Eichner, J. (1969). Discrete conductance fluctuations in lipid bilayer protein membranes, *J. Gen Physiol.* **53**, 741–757.

Cole, K. S. (1949). Dynamic electrical characteristics of the squid axon membrane, *Arch. Sci. Physiol.* **3**, 253–258.

Cole, K. S., and Curtis, H. J. (1939). Electric impedance of the squid giant axon during activity, *J. Gen. Physiol.* **22**, 649–670.

Conti, F., and Neher, E. (1980). Single channel recordings of K^+ currents in squid axons, *Nature* **285**, 140–143.

Ehrenstein, G., Lecar, H., and Nossal, R. (1970). The nature of the negative resistance in bimolecular lipid membranes containing excitability-inducing material, *J. Gen. Physiol* **55**, 119–133.

Ganot, G., Wong, B. S., Binstock, L., and Ehrenstein, G. (1981). In preparation.

Julian, F. J., and Goldman, D. E. (1962). The effects of mechanical stimulation on some electrical properties of axons, *J. Gen. Physiol.* **46**, 297–313.

Loewenstein, W. R., Terzuolo, C. A., and Washizu, Y. (1963). Separation of transducer and impulse generating processes in sensory receptors, *Science* **142**, 1180–1181.

Meuller, P., and D. O. Rudin (1963). Induced excitability in reconstituted cell membrane structure, J. Theor. Biol. **4**, 268–280.

Naitoh, Y., and Eckert, R. (1973). Sensory mechanisms in paramecium. II. Ionic basis of the hyperpolarizing mechanoreceptor potential, *J. Exp. Biol.* **59**, 53–65.

Takeuchi, A., and Takeuchi, N. (1960). On the permeability of end-plate membrane during the action of transmitter, *J. Physiol.* **154**, 52–67.

12

Continuous Stimulation and Threshold of Axons: The Other Legacy of Kenneth Cole

ERIC JAKOBSSON and RITA GUTTMAN

FROM WHENCE WE CAME—HOW KACY COLE AND COLLEAGUES MAPPED THE GEOMETRY OF EXCITABILITY SPACE

In this paper we will discuss an area which is both of high scientific interest and also one in which Kacy Cole has contributed greatly to our understanding. We all know pretty well Kacy Cole's role in clarifying the roles of the membrane impedance and capacitance in excitability culminating in the invention of the voltage clamp (Cole, 1949; Cole and Curtis, 1938), which changed electrophysiology and biophysics drastically and irreversibly. And, as we look in the present-day literature, one is

ERIC JAKOBSSON • Department of Physiology and Biophysics, and Program in Bioengineering, University of Illinois, Urbana, Illinois 61801.
RITA GUTTMAN • Marine Biological Laboratory, Woods Hole, Massachusetts 02543, and Department of Biology, Brooklyn College of the City University of New York, Brooklyn, New York 11210.

tempted to consider the voltage clamp as *the* legacy of Cole, the major landmark that he has left on the scientific landscape. Certainly the clamp has proven to be an indispensable tool for exploring the mechanisms underlying excitability, as is testified to by the army of workers who, armed with clamp, daily do battle with the mysteries of axons, cell bodies, a great array of muscle cells and syncytia, and recently even such things as neuroblastomas (Moolenar and Spector, 1978); but there are some of us, a relatively small fraction of those who clamp, who spend at least part of our time following a somewhat different path in trying to understand excitability, and we find that this path too has first been trod by Kacy Cole. We refer to the study of how ionic currents and the passive properties of nerve combine to produce such properties as thresholds, accommodation, repetitive firing, indeed the whole array of excitable membrane properties that determine how information is organized and integrated in neural membranes. We submit to you the proposition that this area of study has much to tell us about the significance of ionic conductances, and further that it is largely unexplored. If we can talk of such a thing as a scientific landscape, then this area of the integrative properties of excitable membranes is mainly a wilderness, with so far just a few paths cut through the woods, and surprises in store whenever one strays far from the blazed trail. For this is a very tricky landscape to find one's way about in. The perspectives are somehow not what one is used to, and the intuition fails to foresee the effects of such things as saddles, limit cycles, and singular points, all geometrical features of the hyperspace defined by the parameters of the voltage-dependent processes underlying excitability. Or, as we heard one membrane biophysicist (J. W. Moore, personal communication) say one day, "You'd better not say what a set of nonlinear coupled differential equations will do without running it through your computer first."

In this paper, we will first review briefly (not at all comprehensively, owing to lack of space) a selected few of the things known from past work about integrative properties of membranes, emphasizing Kacy Cole's contributions in this area, and then will share with you some results we have recently obtained.

One of the important tools in the theory of integrative membrane properties is simulation by computer. On this occasion it is appropriate to remember that Cole, in collaboration with Antosiewicz and Rabinowitz (Cole, Antosiewicz, and Rabinowitz, 1955), was the first to program the Hodgkin–Huxley (HH) (1952) equations on a digital computer. After FitzHugh and Antosiewicz (1959) worked out a bit of a problem with the singularities in the expressions for the voltage-dependent rate coefficients, two surprising results were seen to emerge from that work. One was that the space-clamped action potential predicted by Hodgkin and Huxley was not all-or-none! If one does with the digital computer what one never could do with the real axon (and never *would* do with a hand calculator!), i.e., vary the stimulus near "threshold" by as little as one part in 10^8, one can generate intermediate size responses. Although not of consequence in predicting the behavior of real squid axons under normal conditions (between 5 and 15°C), this result is of deep importance in understanding the nature of threshold. At higher temperatures this gradedness becomes quite significant. Some years ago, one of us (Guttman) noticed this gradedness in the axons at high temperatures ($>30°$C). Kacy Cole recognized the possible significance of this observation and asked Bezanilla to simulate the HH action potentials at those high temperatures, extrapolating the conductance parameters from low temperature according to the established Q_{10} (Hodgkin and Huxley, 1952). Theory and experiment neatly corroborated each other— with allowances for the distorting effect of the sugar gap (dextrose in this case) (Cole, Guttman, and Bezanilla, 1970). The space-clamped warm squid axon (not to be confused with a "hot" axon, which has large currents) produces responses more like generator potentials than like our normal definition of action potentials. Propagated signals are all-or-none, however. If the threshold gets too broad, the response does not propagate, and no signal is seen far from the stimulus. [When this occurs because of warming, it is the familiar "heat block" (Hodgkin and Katz, 1949).] If the threshold is sufficiently narrow, the response initiated at a point (if the stimulus is above threshold) grows towards some fixed value with distance from the point of stimulus, so that far away from the point of stimulation one sees a signal whose size is independent of the size of the stimulus. Interestingly, although we know that this all-or-none character of the propagated signal is true of the HH axon [because of careful studies by Cooley and Dodge (1966)] and true of real axons, we still lack a general proof that it *must* be true. We do have a neat graphical demonstration by Nasonov (FitzHugh, 1969), which seems particularly compelling for myelinated fibers, but the general proof of the necessity of this phenomenon, of whose necessity we are absolutely convinced, continues to elude us. The missing link is that a theorem about threshold that is obviously true in a two-dimensional space is not necessarily so in a higher-dimensional space. Perhaps we have come up against a mathematical theorem that is true but unprovable.

Earlier in the paper we wrote that the first computer simulations of the Hodgkin–Huxley axon yielded two surprises, but we have not yet mentioned the second. The second is that the model will not accommodate to slowly rising currents without producing action potentials. Of course, no finite number of computer simulations can prove that accommodation to a slowly rising current is impossible, because for any current tried an even more slowly rising one could be postulated. However, the geometrical interpretation of dynamics can come to our rescue in this matter. Setting up a matrix calculation shown in another paper coauthored by Cole (Chandler, FitzHugh, and Cole, 1962), one can calculate the response of the linearized HH model. The rationale behind linearization is that one can then calculate analytically the system's response to a small perturbation. For the HH model, since it has four variables (m, n, h, and V), the solution to its linearized reduction is a sum of four exponentials in time. The arguments of these exponentials yield an important piece of information about the dynamics of the full system. If the real part of any of the four exponential arguments is positive, that means that the stationary point about which the equations were linearized is unstable. An infinitesimal perturbation of the system from that stationary point will not relax back to equilibrium, but will grow. The system can no more remain at such a stationary point (without an external feedback control system, such as a voltage clamp, holding it there) than a pencil can balance itself on a sharpened point.

Numerical results of this matrix calculation are presented graphically by Cooley, Dodge, and Cohen (1965) in a paper prepared, appropriately enough, for a symposium honoring Cole's 65th birthday! It is seen that over a range of the steady-state I–V curve the linearized equations about the points on that curve show instability. The implications are clear and immediate; no matter how slowly a depolarizing current is applied, once the unstable region is entered the system will begin to oscillate, even though there is no negative resistance region in the steady-state I–V curve. This is an amazing result to many people. If one thinks in the time domain, of things changing as one slowly applies the current, one is almost certainly led to reason (as did Hodgkin and Huxley, 1952) that since the Na^+ channels are inactivating and the K^+ channels are opening while the membrane is slowly depolarized, the threshold will rise and stay ahead of the applied current. But intuition misleads in this case, for reasons which are easy to explain in terms of stability theory and perhaps impossible to comprehend in any other way.

WHERE WE ARE—NEW RESULTS ON ACCOMMODATION AND REPETITIVE FIRING

Stability theory leads one to suspect a relationship between accommodation and repetitive firing, since if the singular point is not stable at some point on the I-V curve, there is likely to be a limit cycle at that current, implying repetitive firing. The alternative is for the system to fly off to some infinity somewhere. It is clear on physical grounds that that cannot happen to the Hodgkin–Huxley model (since there are no infinite-voltage batteries in the equivalent circuit) and probably cannot happen to any other realistic nerve model either. So we know that lack of accommodation (an unstable region on the I-V curve) implies repetitive firing. The converse is not necessarily true. It is possible to have, at some value of applied current, both a limit cycle and a stable singular point. So repetitive firing does not *necessarily* imply lack of accommodation—but it might. It occurred to us to ask some squid axons. We knew that squid axons can be made to fire repetitively by lowering extracellular divalent cations—would axons in low divalent cations accommodate to slowly rising currents? The answer is given in Figure 1 (from Jakobsson and Guttman, 1980). In Figure 1b is an oscilloscope trace of a typical experiment in which slowly rising currents are applied to a squid axon in 10 mM calcium and no magnesium. Time scale is 100 msec/div. One can see that even for a ramp current applied so slowly that essentially nothing happens for almost a second, the axon finally breaks into repetitive activity, rather than accommodating. Note that these experiments were done with a sucrose gap arrangement, so that in addition to the normal ionic currents, the system equivalent circuit would include the extra hyperpolarizing current driven by the emf generated by the sucrose–saline interface (Julian, Moore, and Goldman, 1962).

The curves in Figure 1a are the voltages predicted from very slow ramp currents applied to the normal Hodgkin–Huxley axon. The qualitative similarity of the two sets of curves is remarkable, especially because of the differences we know of in the details of the two cases. The

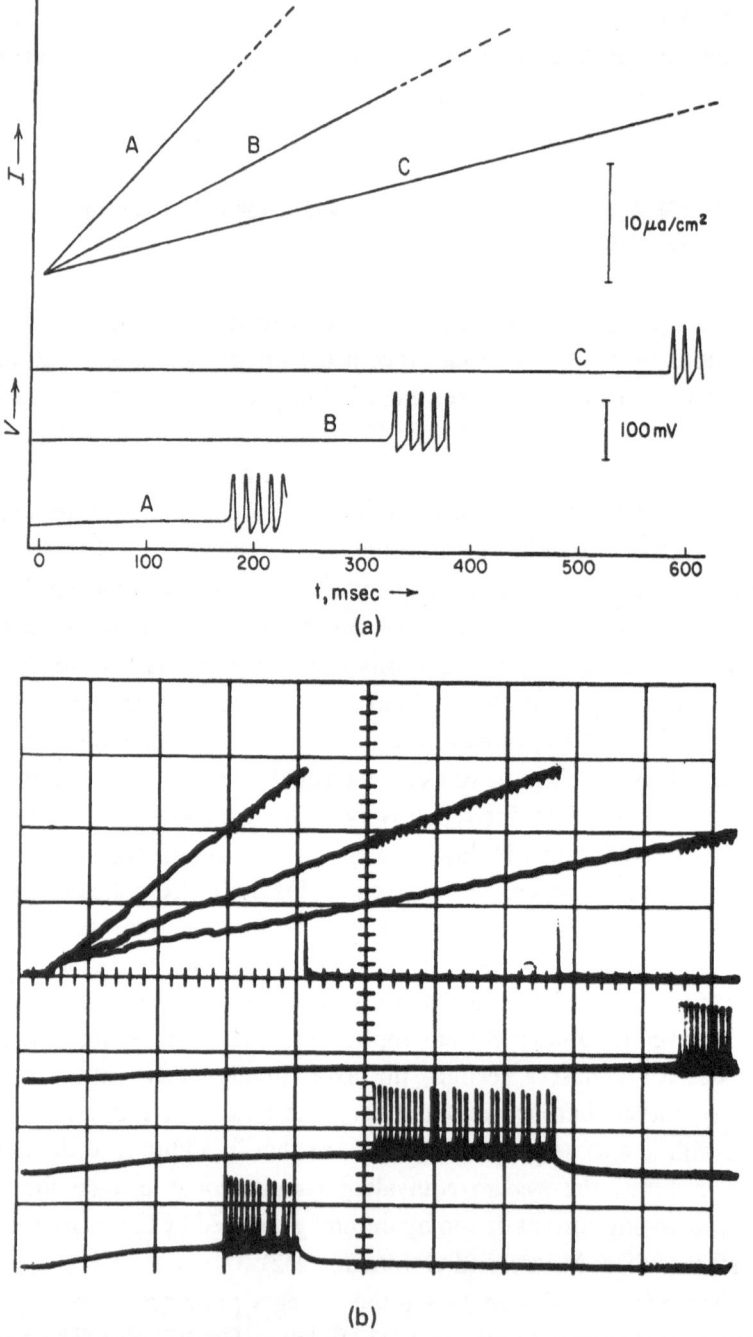

Figure 1. Lack of accommodation by Hodgkin–Huxley model axon (a) and actual squid axon in reduced divalent cation concentration (b), in response to slowly applied depolarizing current. From Jakobsson and Guttman (1980).

experiments were run in low divalent cation concentration, whereas the model parameters are appropriate for normal divalent cations, and the model has no provision for the sucrose gap hyperpolarizing current, nor for the effects of K^+ accumulation in the periaxonal space (Adelman and FitzHugh, 1975). Because of these differences (and possibly other reasons as well), the magnitude of the stimulating current density to elicit a given behavior is much greater for the experiments than for the simulations. But the essential fact of nonaccommodation is shared by both real squid axon and the Hodgkin–Huxley model, as well as the qualitative shapes of the curves, suggesting that the qualitative features of their phase spaces (stability of singular points, existence of limit cycles, etc.) are the same. One aspect of these results was a surprise to us. Close examination of the curves reveals a small but definite tendency for the repetitive firing to start at a lower current when the current is applied more slowly. This is not just lack of accommodation but in fact its opposite; the threshold apparently gets *lower* as the stimulus is maintained. We were initially rather skeptical of this experimental result. Even though it is consistent from axon to axon, it just seemed not right, somehow—but then we found that the HH axon does the same thing. And the magnitude of the effect, as can be seen from Figure 1, is about the same. These experiments were done near the end of the last experimental season that we were able to work at Woods Hole, so unfortunately we have no further experimental results in hand relating to this phenomenon. We have been able to explore the model's behavior further, however. Figure 2 shows the results of computer runs on the HH axon with ramps of quite a few different slopes, where we plot the current at which firing is initiated versus the slope. More surprises. Instead of a smooth curve, we get something like a sawtooth. The apparent threshold current "jitters." Previously we had been skeptical of the experiments; now we got suspicious of the computer. From previous experience with simulating sinusoidal current stimulation of axons (Guttman, Feldman, and Jakobsson, 1980) we knew that threshold phenomena were sometimes a severe test of numerical integration techniques (also pointed out by Moore and Ramon, 1974), and that sometimes "noise" generated by errors in numerical integration could give peculiar threshold behavior which was spurious, in the sense that it was not an accurate representation of the "true" behavior of the HH model (although such "spurious" behavior might be actually a *better* representation of the real axon, which,

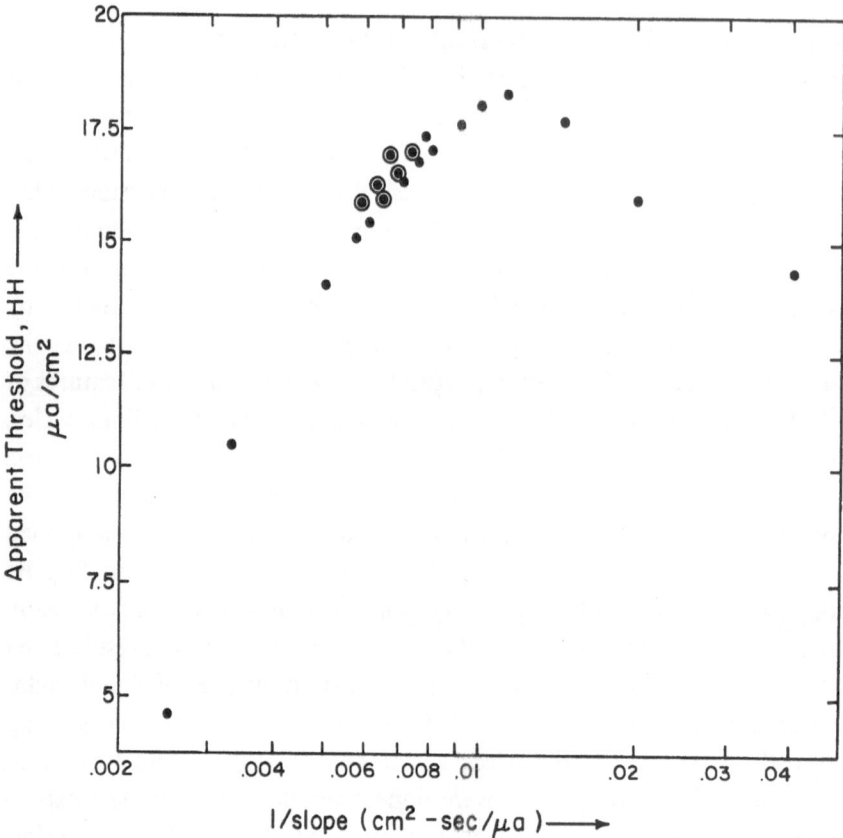

Figure 2. For standard Hodgkin–Huxley model axon at 6.3°C. For slowly applied ramp currents starting from rest (−60 mV), time at which action potentials are first elicited versus reciprocal of ramp slope. Computations done with CSMP-3 program on IBM 360/75 (Speckhart and Green, 1976). Solid symbols were computed with Gear method for stiff differential equations. Open circles surrounding solid symbols represent computations repeated with fourth-order Runge–Kutta method with variable step size. The two methods gave essentially identical results in each case.

unlike the HH axon, is noisy). In our experience, the generally least noisy numerical integration technique has been the Gear method for stiff sets of differential equations and it is the one we routinely use now for threshold computations. On seeing the results in Figure 2, however, we redid several of the runs in the suspect region using a fourth-order Runge–Kutta method with variable step size, on grounds that if we had an accuracy problem, we would not get the *same* inaccurate result for two different integration techniques. The circles represent runs done with the Runge–Kutta method as well as the Gear method, and it is seen that the results of the Gear method were precisely duplicated by the Runge–Kutta method. So, we concluded that our computer was not lying to us after all, and that it was incumbent on us to try to understand what the machine was trying to tell us.

Figure 3 shows these results in more detail. Here we have expanded the section of the previous graph that gave us the "funny" results and

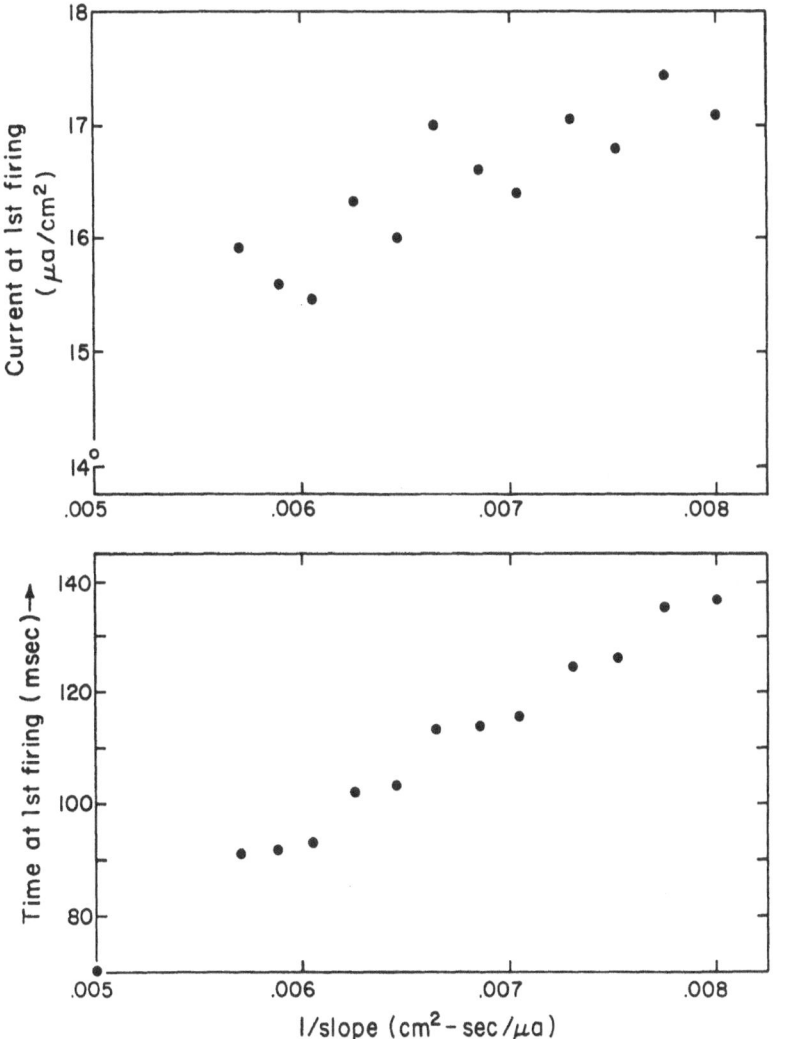

Figure 3. On expanded scale, current at time of first action potential versus reciprocal of ramp slope (upper), and time of first action potential versus reciprocal of ramp slope (lower), for standard HH axon with ramp applied from rest (−60 mV) at time = 0.

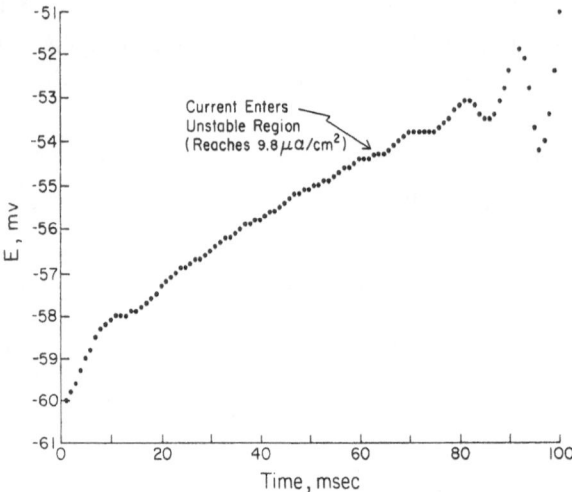

Figure 4. Computer output of first 100 msec of voltage response to a ramp current of 155 μA/cm^2 sec applied from rest. It is seen that subthreshold oscillations begin shortly after the current enters the "unstable region" (9.8 μA/cm^2), but that a substantial delay occurs before action potentials begin. In this instance, the first action potential occurred on the very next oscillation after the last one pictured, at 102 msec.

plotted time to first action potential as well as current. A pattern begins to emerge, in which the time goes up in a series of steps, and the "sawtooth" pattern of the apparent current threshold (upper portion of Figure 3) is clearly related to the plateaus in the "staircase" pattern of the time-to-fire graph (lower portion of Figure 3). The key to understanding the system's behavior is present in this pattern, and is made explicit in Figure 4. Here is shown the response to a slope current of 155 μA/cm^2 sec. Note that sometime before the action potentials start one sees distinct subthreshold oscillations which grow to a millivolt or two amplitude with a frequency of about 100 Hz.* The action potentials always start off the top of one of the subthreshold oscillations, explaining why the time-to-first-action potential is a "staircase" function of ramp current slope, and why the apparent current threshold is a sawtooth rather than smooth function of ramp current slope (Figure 3).

The position on the I-V curve above which the singular points are unstable is 9.8 μA/cm^2. If one can assume that very minute oscillations

*This frequency is close to those calculated from the linearized HH equations at rest and critical I [Rinzel, J. (1978). *Fed. Proc.* **37**, 2793–2802].

begin when the value of applied current crosses that value, then there is quite a long delay from the time the oscillations begin until they develop into action potentials. This delay explains the phenomenon of "reverse accommodation." For very slowly rising currents, the current will not reach a value much higher than 9.8 μA/cm^2 by the time the action potentials start.

So the essential phenomenon we are looking at here, which underlies the apparent "reverse accommodation" (at least in the model), is a delay. The model membrane does not always quickly produce action potentials when subjected to a sufficient stimulating current, but may wait for quite a long time before commencing. How long? Perhaps an infinite or arbitrarily long time, if one were to slowly take the current up to the edge of the unstable region in the I–V curve, and then hold it steady. In the slowest ramp previously shown in Figure 1, of a ramp current of 25 μA/cm^2 sec, the current passed the 9.8 μA/cm^2 mark at $t=392$ msec but the first action potential did not come until 576 msec, a delay of 184 msec between the time the applied current entered the unstable region and the time of the first action potential. This delay is shown explicitly in Figure 5. An even longer delay can be achieved by slowly raising the current into the unstable region and then holding it constant. For example, we did a simulation in which we ramped the current up to 10 μA/cm^2 at a rate of 100 μA/cm^2 sec and then held the current at 10

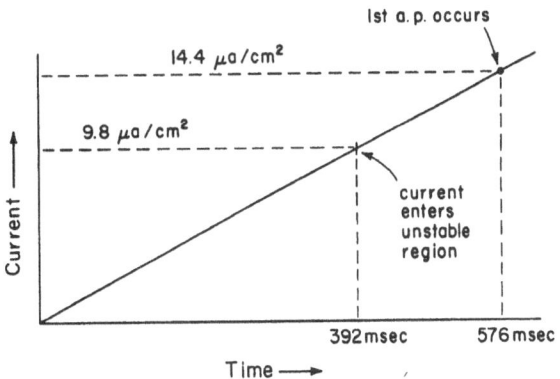

Figure 5. Graphical representation of delay for most slowly applied ramp current of Figure 2. When ramp current was applied to HH axon at 0.025 μA/msec cm^2, the unstable region was entered at 392 msec, but action potentials did not begin until 576 msec, a delay of 184 msec.

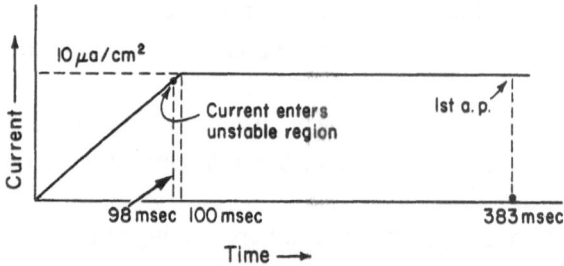

Figure 6. Graphical representation of HH axon delay for current slowly ramped into the unstable region and held at a current just barely in that region. Current is ramped up to $10\,\mu A/cm^2$ at a rate of $0.01\,\mu A/msec\,cm^2$ and then held at $10\,\mu A/cm^2$. Current enters unstable region at 98 msec but action potentials do not begin until 383 msec, a delay of 285 msec.

$\mu A/cm^2$. The results of this stimulation are shown in Figure 6. It is seen that there is a 285-msec delay in this instance between the time of entering the unstable region and the time of first action potentials.

WHITHER WE GO—WHAT MIGHT "REVERSE ACCOMMODATION" MEAN FOR NEURAL CODING?

The take-home lesson from all this is that there may be a very long delay indeed (several hundred milliseconds) between a stimulus and a response in a relatively simple excitable membrane described by standard Hodgkin–Huxley conductances. This delay can result in such a peculiarity as an apparent "reverse accommodation," and perhaps other complicated dependences of response on temporal course of stimulus, not at all well described by the traditional concepts of accommodation which appear in textbooks. Speaking very speculatively, it seems to us likely that such delays are significant in neural coding. We have no evidence for that, but when one is aware of how useful delays are in design of electronic circuits, it seems unlikely that Nature would neglect to exploit them in the design of nervous systems. And of course, insight into coding is a major scientific justification for doing studies of threshold properties in the first place.

We would like to close with a graphic summary, shown in Figure 7, of where we are scientifically and how Kenneth Cole has led us there. On the flow chart of progress in axon electrophysiology we have shown the

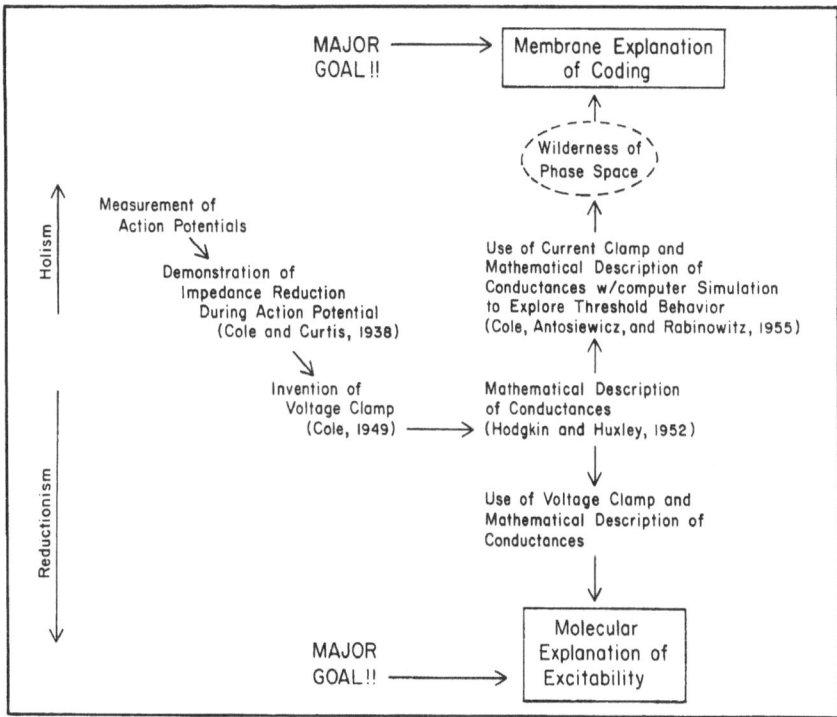

Figure 7. Rough chart of progress in reductionist and holistic directions in axonology. Events in which Kenneth Cole has played a particularly leading role are noted. We suggest that progress up to the Hodgkin–Huxley axon was primarily reductionist, culminating in the mathematical description of the conductance changes underlying the action potential. The HH axon has provided a reference point for further progress in both the reductionist and holistic directions, as it has provided a framework for interpreting voltage clamp data and a means for simulating spike generation. In this paper we have followed Cole's lead in using computer simulation of the Hodgkin–Huxley axon to explore threshold phenomenon in nerve.

original reductionist thrust that led from the accurate measurement of resting and action potentials through the dramatic demonstration of impedance reduction during the action potential (Cole and Curtis, 1938, which produced, as a by-product, one of the most esthetically pleasing shapes in all of physiology, i.e., the oscilloscope trace which graces the Cole Medal) to the voltage clamp technique (Cole, 1949) culminating in the construction of the Hodgkin–Huxley model. After Hodgkin and Huxley, the path of progress bifurcates. The model has been an important tool in correlating voltage clamp data, leading us farther down the reductionist path towards the major goal of a molecular explanation of excitability. At the same time, with Cole's leadership, the model has enabled us to start back in the holistic direction, through the wilderness of phase space, towards another major goal of electrophysiology—an understanding of coding in terms of ionic conductances. In this paper we hope to have contributed in at least a small way to widening and straightening the trail through phase space which Kacy Cole began to blaze for us some time ago.

ACKNOWLEDGMENTS

Support was received from the Research Board, the Bioengineering Program, and the Department of Physiology and Biophysics of the University of Illinois, and by grant No. HEW PHS RO1 NS 12272-03 from the National Institutes of Health. Technical assistance for the experiments was given by Stephen Lewis. Dr. John Rinzel made helpful comments on the manuscript.

REFERENCES

Adelman, W. J., Jr., and FitzHugh, R. (1975). Solutions of the Hodgkin–Huxley equations modified for potassium accumulation in a periaxonal space, *Fed. Proc.* **34**, 1322–1329.

Chandler, W. K., FitzHugh, R., and Cole, K. S. (1962). Theoretical stability properties of a space-clamped axon, *Biophys. J.* **2**, 105–127.

Cole, K. S. (1949). Dynamic electrical characteristics of the squid axon membrane, *Arch. Sci. Physiol.* **3**, 253–258.

Cole, K. S., Antosiewicz, H. A., and Rabinowitz, P. (1955). Automatic computation of nerve excitation, *J. Soc. Indust. Appl. Math.* **3**, 153–172.

Cole, K. S., and Curtis, H. J. (1938). Electric impedance of nerve during activity, *Nature* **142**, 209.

Cole, K. S., Guttman, R., and Bezanilla, F. (1970). Nerve membrane excitation without threshold, *Proc. Nat. Acad. Sci. U.S.A.* **65**, 884–891.

Cooley, J. W., Dodge, F. A., Jr. (1966). Digital computer solutions for excitation and propagation of the nerve impulse, *Biophys. J.* **6**, 583–599.

Cooley, J., Dodge, F., and Cohen, H. (1965). Digital computer solutions for excitable membrane models, *J. Cell. Comp. Physiol.*, Suppl. 2, Pt. II **66**, 99–109.

FitzHugh, R. (1969). Mathematical models of excitation and propagation in nerve, in *Biological Engineering*, H. P. Schwan, Ed. (McGraw-Hill, New York), pp. 1–85.

FitzHugh, R., and Antosiewicz, H. A. (1959). Automatic computation of nerve excitation—detailed corrections and additions, *J. Soc. Indust. Appl. Math.* **7**, 447–458.

Guttman, R., Feldman, L., and Jakobsson, E. (1980). Frequency entrainment of squid axon membrane, *J. Membr. Biol.* **56**, 9–18.

Hodgkin, A. L., and Huxley, A. F. (1952). A quantitative description of membrane current and its application to conduction and excitation in nerve, *J. Physiol.* **117**, 500–544.

Hodgkin, A. L., and Katz, B. (1949). The effect of temperature on the electrical activity of the giant axon of the squid, *J. Physiol.* **109**, 240–249.

Jakobsson, E., and Guttman, R. (1980). The standard Hodgkin–Huxley model and squid axons in reduced external Ca^{++} fail to accommodate to slowly rising currents, *Biophys. J.* **31**, 293–297.

Julian, F. J., Moore, J. W., and Goldman, D. E. (1962). Membrane potentials of the lobster giant axon obtained by use of the sucrose-gap technique, *J. Gen. Physiol.* **45**, 1195–1216.

Moolenar, W. H., and Spector, I. (1978). Ionic currents in cultured mouse neuroblastoma cells under voltage-clamp conditions, *J. Physiol.* **278**, 265–286.

Speckhart, F. H., and Green, W. L. (1976). *A Guide to Using CSMP—The Continuous System Modeling Program* (Prentice-Hall, Englewood Cliffs, New Jersey).

13

A Model of Drug-Channel Interaction
in Squid Axon Membrane

DANIEL L. GILBERT and RAYMOND J. LIPICKY

For some drugs which inhibit sodium conductance, the extent of inhibition is highly dependent upon the recent history, or "use," of the sodium channels (i.e., recent depolarization of the membrane). Because of the relationship to sodium channel "use" this phenomenon has become known as a use-dependent effect. Drugs which exhibit use-dependent

DANIEL L. GILBERT • Marine Biological Laboratory, Woods Hole, Massachusetts 02543, and Laboratory of Biophysics, IRP, National Institute of Neurological and Communicative Disorders and Stroke, National Institutes of Health, Bethesda, Maryland 20205. RAYMOND J. LIPICKY • Division of Clinical Pharmacology, Department of Medicine and Department of Pharmacology and Cell Biophysics, University of Cincinnati Medical Center, Cincinnati, Ohio 45267; Division of Drug Biology (HFD-410), Food and Drug Administration, 200 C Street, S.W., Washington, D.C. 20204; Division of Cardiorenal Drug Products (HFD-110), Food and Drug Administration, 5600 Fishers Lane, Rockville, Maryland 20857; Marine Biological Laboratory, Woods Hole, Massachusetts 02543; and Laboratory of Biophysics, IRP, National Institute of Neurological and Communicative Disorders and Stroke, National Institutes of Health, Bethesda, Maryland 20205.

effects also inhibit sodium conductance when there has been no recent "use." This effect has come to be known as the tonic effect. A number of investigations primarily concerned with use-dependent block of Na channels have been reported (Strichartz, 1973; Courtney, 1975; Yeh and Narahashi, 1976; Khodorov et al, 1976; Hille, 1977; Schwarz et al., 1977; Lipicky et al., 1978; Cahalan, 1978; Yeh, 1978; Cahalan and Almers, 1979). In these investigations some local anesthetics and their derivatives (mainly QX-314 and GEA 968), benzocaine, quinidine and yohimbine have been studied in squid axon, myelinated frog nerve, and frog skeletal muscle. In addition several other chemicals have been shown to have use-dependent effects on sodium channels such as paragracine (Seyama et al., 1980), strychnine (Shapiro et al., 1974; Shapiro, 1977), and pancuronium (Yeh and Narahashi, 1977).

Variables that have been shown to be important to the use-dependent phenomenon (at least in some cases) have included (a) the charge on the molecule; (b) transmembrane potential; (c) the direction of currents (i.e., inward or outward across the membrane); (d) the duration of depolarizing pulses; (e) the frequency of depolarizing pulses; (f) the state of (or effects of the drug upon) sodium inactivation; (g) the state of effects (or) of the drug upon slow sodium inactivation; and/or (h) the relative binding of drug to open versus closed channels. These variables have been invoked in a variety of ways and in various combinations to explain the mechanisms of action of the particular drug, or class of drugs, under study.

The eventual result, regardless of the specific combination of the above variables, is generally assumed to be the transformation of a conducting channel into a non-conducting channel; and in that sense the end result is channel occlusion (i.e., simple blockage). A simple occlusive mechanism is strongly suggested for QX-222 in the frog neuromuscular junction (Neher and Steinbach, 1978) as determined by recording from single channels; however, these channels are chemically sensitive. Occlusion of voltage-dependent sodium channels by drugs other than tetrodotoxin (TTX) and saxitoxin has not yet been directly verified in axonal membranes.

We have proposed a nonocclusive mechanism (i.e., a modified kinetic model) for yohimbine action (Lipicky et al., 1978), which remains a plausible mechanism. In this communication we intend to do the following:

- a. present our modified kinetic model in somewhat more detail;
- b. present some additional descriptive information that further characterizes the effects of yohimbine in the squid axon, and supports our model;
- c. present observations on QX-314, in squid axon, which also support the modified kinetic model we proposed for yohimbine; and moreover suggest that this could be a mechanism for some drugs other than yohimbine.

None of our observations prove the validity of the model. There are, however, sufficient data consistent with it to make it a reasonable alternative to other models that have been presented in an attempt to account for use dependence.

MODIFIED KINETIC MODEL

Since this symposium is in honor of Kenneth S. Cole, who appreciated the use of models in explaining biophysical phenomena (Cole, 1965), we will begin with a description of our model.

Drugs can conceivably affect ionic channels either by blocking the channels or by altering the kinetics of the channels, or by a combination of both these actions. Figure 1 shows a general scheme for any kinetic model. A channel can exist either in the free state without any drug binding or in the drug-bound state. The normal, free state can either be in the closed configuration (C) or in the open configuration (O). Likewise, the drug-bound state can also be in the closed (N) or open (D) configurations. For either free or drug-bound channels the closed state includes the inactivated state. Thus, we need deal with only four states: the free-closed state, the free-open state, the drug-bound-closed state, and the drug-bound-open state. In this model the free-closed state does not pass directly into the drug-bound-open state; rather the channel must pass through an intermediate state, i.e., either the free-open state and/or the drug-bound-closed state. Likewise, the free-open state must proceed via two steps to get to a drug-bound-closed state. In other words, we have made the simple assumption that the binding–unbinding process and the opening–closing process cannot occur simultaneously. That is, if both

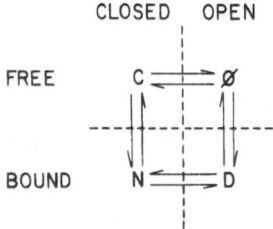

Figure 1. General schematic for drug effect model. Channels can be free (i.e., not drug-bound) and exist in the closed (*C*) or open (*O*) conformation; or they can be drug-bound and exist in the closed (*N*) or open (*D*) conformation. The closed state includes the inactivated state.

processses occur, then one of these processes must precede the other. In the general model, the drug can bind to a channel when the channel is in either the open or the closed configuration.

We assume that a channel in the free-closed state undergoes a conformational change to become a free-open channel. This being a normal channel, the kinetics of the channel are assumed to be identical to those described by Hodgkin and Huxley (1952). The channel in the drug-bound-closed state also undergoes a conformational change to become a drug-bound-open channel. The kinetics of this reaction could have any kind of time and/or voltage dependence; for our convenience we have assumed a simple modification of the kinetics described by Hodgkin and Huxley. When in the open state, whether free or drug bound, the channel can conduct.

For a simple cyclic scheme, as illustrated in Figure 2, we see that for a four-state model, there are four equilibrium constants and eight rate constants. The product of the equilibrium constants must equal unity. Since the equilibrium constant is the forward rate constant divided by the backward rate constant, the product of the forward rate constants divided by the product of the backward rate constants is also unity. Or the product of the forward rate constants (k_1, k_3, k_5, and k_7 in Figure 2)

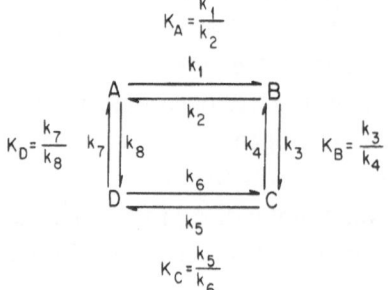

Figure 2. Schematic diagram for a general, cyclic, kinetic reaction. For this figure only, forward rate constants are clockwise and backward rate constants are counterclockwise.

equals the product of the backward rate constants (k_2, k_4, k_6, k_8 in Figure 2). These relationships determine a number of features of the model:

(a) Even if there is no direct path between A and D, there is a path from A to D via B and C. The ratio of A to D is fixed, even though there may be no direct connection. The reciprocal of the equilibrium constant of A with respect to D equals the product of the other equilibrium constants, or equals the product of the backward rate constants divided by the product of the forward rate constants.

(b) Note the similarity between Figures 1 and 2 (i.e., $A=C$, $B=0$, $C=D$, and $D=N$). For Figure 1, if we define P as the ratio of free-open to free-closed channels at a given instant and at a designated potential, then P is a function of the Hodgkin–Huxley parameters. P is therefore a coefficient whose value changes with time and potential. If we define Q as the ratio of drug-bound-open to drug-bound-closed channels, then similarly Q is a function of the modified Hodgkin–Huxley kinetics. These two relations can be expressed in the following:

$$O = P \cdot C \tag{1}$$

$$D = Q \cdot N \tag{2}$$

If we further compare Figure 1 and Figure 2 by defining k_3 as the forward rate coefficient for the binding of the drug to the open channels and k_4 as the backward rate constant, then K_B, the binding constant is defined as

$$K_B = k_3 / k_4 \tag{3}$$

Likewise, we can define k_8 as the forward rate coefficient for the dissociation of drug-bound channels to free channels and k_7 as the backward rate constant, then K_D, the dissociation constant, is defined as

$$K_D = k_7 / k_8 \tag{4}$$

Note that also

$$K_B = O_e / D_e \tag{5}$$

where the subscript e refers to an equilibrium state. Similarly,

$$K_D = C_e / N_e \tag{6}$$

It follows therefore that

$$K_B \cdot K_D = Q_e / P_e \tag{7}$$

(c) The ratio Q_e/P_e is an equilibrium value and is not a function of time. However, it is a function of potential because Q is a function of the modified Hodgkin–Huxley kinetics. This provides a way of distinguishing free channels from drug-bound channels by the method of measuring the total ionic conductances. Since the ratio Q_e/P_e is a function of potential, it follows that the ratio K_B/K_D is also a function of potential.

In other words, the binding and the dissociation constants for drugs, in the model, are dependent upon potential. Equation (7) for our drug model is simply a restatement that the product of the equilibrium constants equals one for a cyclic reaction. Since the ratio K_B/K_D is determined by the four rate constants for binding and unbinding, then at least one of these rate constants is voltage dependent. If the binding does not occur in the open state, then there are only two rate constants, and at least one of these is voltage dependent. If the binding only occurs in the open state, then at least one rate constant is voltage dependent.

In our model, we have assumed for simplicity that the drug-bound sodium channels open and close according to a simple modification of Hodgkin–Huxley kinetics for sodium. For drug-bound sodium channels, we have modified each of the Hodgkin–Huxley (1952) rate constants by multiplying one or more parameters (i.e., α_m, β_m, α_h, or β_h) by a constant factor which we have defined as the drug amplification factor. There are two drug amplification factors for the m or activation process and two drug amplification factors for the h or inactivation process. If the drug amplification factor equals unity, then the rate constant for the bound channels is the same as the Hodgkin–Huxley value (i.e., it is not modified).

The results of a computer simulation are shown in Figures 3 and 4. For this example, a rest and holding potential of -65 mV were chosen. A dose of the drug was chosen such that after a steady state was reached, 30% of the channels were in a drug-bound state (approximately equivalent to a 30% tonic effect; approximate because drug-bound-open channels conduct and one measures total current). A 60-mV step pulse lasting

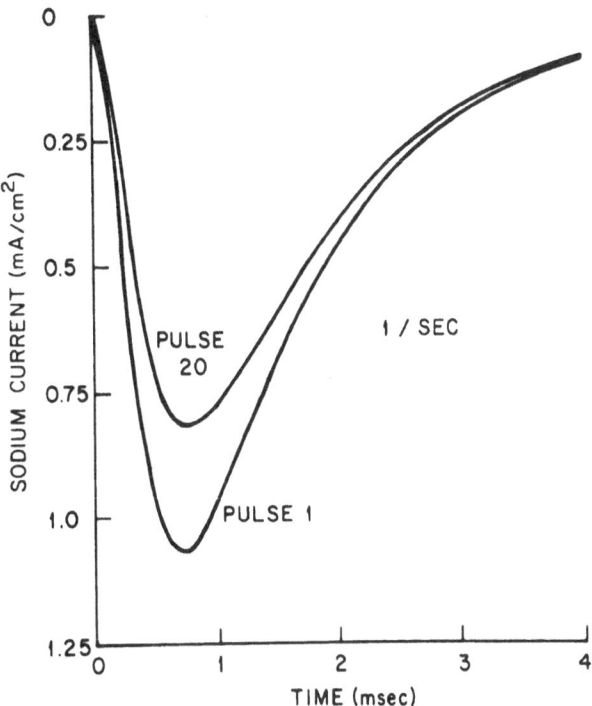

Figure 3. Computer simulated current–time plots for the first and 20th pulse of a 1 per second train. Conditions were: holding potential $= -65$ mV, pulse potential $= 60$ mV (absolute membrane potential -5 mV), pulse duration $= 4$ msec, free-open to drug-bound-open channel rate coefficient $= 0.5$ drug concentration units/msec, initial ratio of drug-bound to total channels (tonic effect) $= 30\%$, steady-state ratio of drug-bound to total channels during a pulse (use-dependent effect) $= 80\%$. The peak sodium conductance, for a given step potential, divided by the maximum sodium conductance is equivalent to the fraction of sodium channels open at the time the peak conductance was obtained. If the unit conductance of a drug-bound-open channel is the same as that of a free-open channel, then it is possible to convert the percentage of open channels at any time to the peak sodium conductance or sodium current. For the conversion of percent open channels to current, it was assumed that E_{Na} was $+50$ mV and $\bar{g}_{Na}$ was 120 mS/cm^2. Thus, if there were no drug-bound channels and assuming HH kinetics, the maximum sodium current for this pulse would be 1.46 mA/cm^2, which would correspond to 22.1% of the total channels being in the open state.

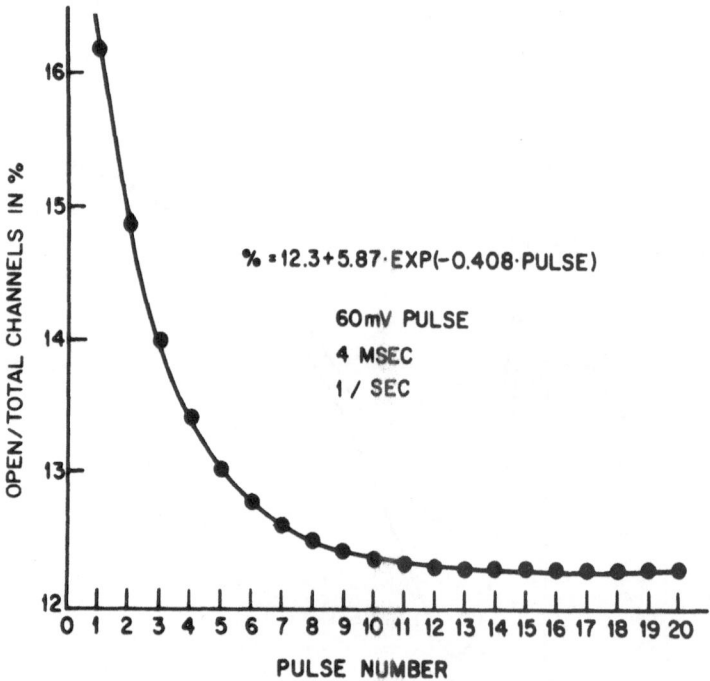

Figure 4. Observed use-dependent effect of the model depicted in Figure 2, with parameters as listed in Figure 3. The maximum sodium current (reflecting the total-open channels) is plotted as a function of pulse number. Since the pulsing rate was 1 per second, the pulse number is equivalent to time in seconds and zero time is at pulse number 1. The line through the points is a single exponential, as determined by a least-squares fit.

4 msec was applied at a repetition rate of 1 per second. At this pulse potential, the absolute membrane potential is -5 mV. The model requires that if this pulse were to be applied for an infinite period of time, at this potential 80% of the channels would eventually be in a drug-bound state (approximately equivalent to an 80% total inhibition, or a 71% $[(70-20)/70]\times100$ use-dependent effect). It was also assumed that the free-open to drug-bound-open channel, unidirectional, rate coefficient (k_3) was equal to 0.5 drug concentration units/msec and that there was no direct interchange between the free-closed and drug-bound-closed channels. The drug amplification factors for the α_h, β_h, and the β_m rate constants were all set equal to unity. The drug amplification factor for α_m was set to 0.3 both for the rest and pulse potentials. The channel states were calculated every 0.01 msec.

For these conditions, the drug-bound-open to free-open channel rate constant (k_4) was calculated to be 38.6 msec^{-1} during the rest condition (which lasted 996 msec for each cycle), and 0.159 msec^{-1} during the pulse condition (which lasted 4 msec for each cycle). The backward rate constant (drug unbinding, k_7) was two orders of magnitude smaller (i.e., more drug binding), during the pulse. For this particular simulation the binding rate coefficient (k_8) was fixed by preselection and k_7 was calculated. The convention could have been opposite and the simulation would have had an identical result (the use of a particular rate constant that exhibits voltage dependent properties does not alter the results of the simulation, as only one rate constant need be voltage dependent). By calculation, during the initial rest condition, there were 0.006% channels in the drug-bound-open state, and 69.994% were in the free-closed state. Thus the initial 30% drug-bound channels are almost entirely drug-bound-closed channels.

Figure 3 shows the currents calculated for pulses 1 and 20, plotted as a function of time. Figure 4 shows the maximum number of open channels (free-open plus drug-bound-open) for each pulse. One can obtain typical use dependence simply by assuming the following: more drug binding with channels in the open conformation and the drug affecting only the parameter α_m. More complex drug-bound channel kinetic modification (provided it were in the direction of slower or voltage dependence more in the depolarizing direction) would result in more use dependence.

Of some importance is the shape of the current–time plots shown in Figure 3. The twentieth pulse, compared to the first pulse, has an apparent mildly decreased rate of inactivation (i.e., the rate of return toward zero current, after the peak current). This is also the case for current–time records published for yohimbine (Lipicky et al., 1978), QX-314 (Strichartz, 1973), and 9-amino acridine (Cahalan, 1978). In the simulation, this apparent decrease in rate of inactivation results from the contribution of drug-bound-open channels to the total current. The time course of current from drug-bound-open and from free-open channels, as well as the total current, for this single simulation is shown in Figure 5 for illustration of this point.

In the simulation, if initial conditions were set so that at zero time all channels were in the free state (i.e., not yet exposed to drug), the time course of development of the 30% drug-bound channel state (i.e., the

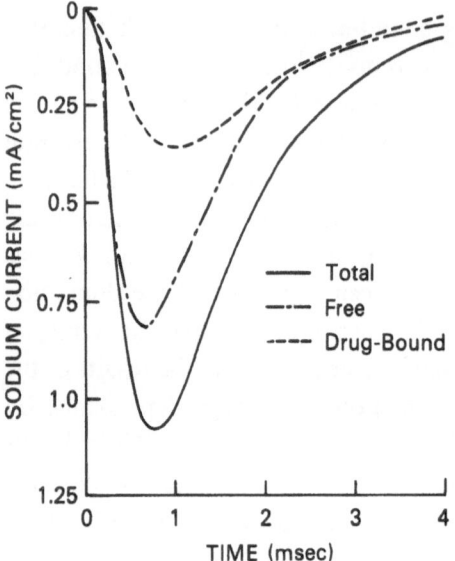

Figure 5. Computer simulated current–time plots for total current (drug-bound-open and free-open channels), free-channel current (free-open channel) and drug-bound channel current (drug-bound-open channels). Conditions were the same as for Figure 3. The results shown are currents from the first pulse of a 1 per second train.

time course of the development of the tonic effect) could be determined. All conditions were as specified above, and of particular note, there was no exchange between drug-bound-closed and free-closed channels (rate constants k_7 and k_8 were not present in the equations) during the simulation. The time course of development of drug-bound channels (as stated above, these consist mainly of drug-bound-closed channels) was exponential and when fitted by a least squares procedure has a time constant of 6.8 sec. The dependence of the time course of development of the tonic effect on the free-open to drug-bound-open rate constant is shown in Table I.

Thus when the drug is first applied to the resting membrane, a portion of the channels will become bound to the drug, even if all drug binding only occurs when channels are in the open state. This is because

Table I. Relationship of Time Constant of Development of Tonic Effect to the Free-Open to Drug-Bound-Open Rate Coefficient

Rate Coefficient[a]	Time constant[b]	Rate coefficient[a]	Time constant[b]
10	0.34	0.5	6.8
5	0.68	0.1	34.0
1	3.4	0.01	310.0

[a]Free-open to drug-bound-open rate coefficient (drug concentration units/msec.
[b]Time constant of development of drug-bound-closed channels (sec).

there are always some open channels, even if they are small in number. Now, if the membrane potential is changed, there will be a new equilibrium proportion of drug-bound channels to total channels [this is true because of voltage dependence of the drug binding rate constants as defined in part (c), discussed above]. This change in the equilibrium proportion of the drug-bound channels increases the proportion of drug-bound channels during a step change in potential, during voltage clamp. If the step is terminated before the new equilibrium value is reached, and the membrane potential is returned to its resting value, the previous condition will produce a decrease in the drug-bound channels with time. Thus, after the step pulse is ended, the number of drug-bound channels at the beginning of the rest period will be at some value between the rest equilibrium value and the pulse equilibrium value, and during a long rest period, the number of drug-bound channels will eventually return to its original equilibrium rest value. However, if another step pulse occurs before the equilibrium rest value is reached, then at the start of the step pulse, the number of drug-bound channels will again increase to some value between the equilibrium rest value and the equilibrium pulse value. When a pulse of the same magnitude and duration is applied at a constant rate, a steady state will eventually be reached in which,

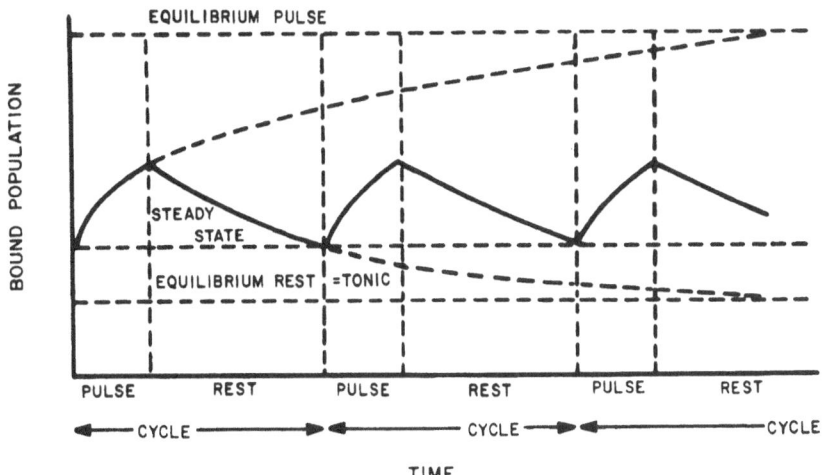

Figure 6. Schematic of the changes in drug-bound channel populations after steady-state use dependence has developed during repetitive pulsing. Consult text for details. Shapes are qualitative in the figure (i.e., for sense and direction only) and are not reproductions of a simulation.

during any designated time in any given cycle, the number of drug-bound channels will remain the same.

This sequence of events is shown in Figure 6, which is drawn schematically for diagramatic purposes, and is not a product of any simulation. During the pulse, the rate of formation of the drug-bound channels will depend upon (a) the rate constants for drug-binding and unbinding, (b) the Hodgkin–Huxley parameters for the free channels, and (c) the analogous parameters for the drug-bound channels. The shape of the rise in drug-bound channels depicted during a step pulse, in Figure 6, may not be an accurate form for the rate of increase in drug-bound channels; rather, it is given only to show the directional changes for the initial and final drug-bound channels during step pulses after a steady state is attained.

YOHIMBINE

Yohimbine (Figure 7) is a naturally occurring indolealkylamine alkaloid that is best known for its alpha-adrenergic blocking properties (Hamet, 1925; Nickerson, 1949; Werner, 1958). Carbons 3, 15, and 20 are asymmetric centers which give rise to four absolute configurations (normal, pseudo-, allo-, and epiallo-) for the molecule. For each config- uration, side groups on carbons 3, 15, 16, and 17 can exist in *cis* or *trans*

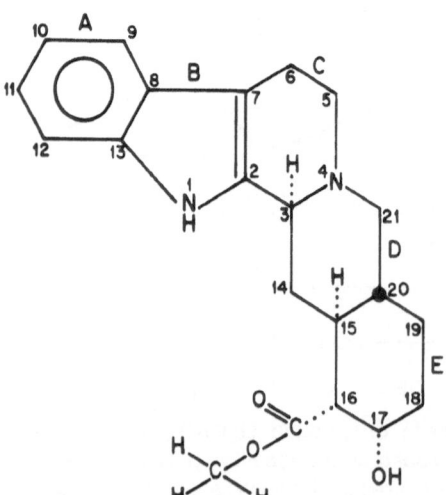

Figure 7. Structural formula for yohimbine. Side groups can be *cis* or *trans* about carbons 3, 15, 16, and 17. Carbons 3, 15, and 20 are asymmetric and give rise to the possible four different absolute configurations of the molecule.

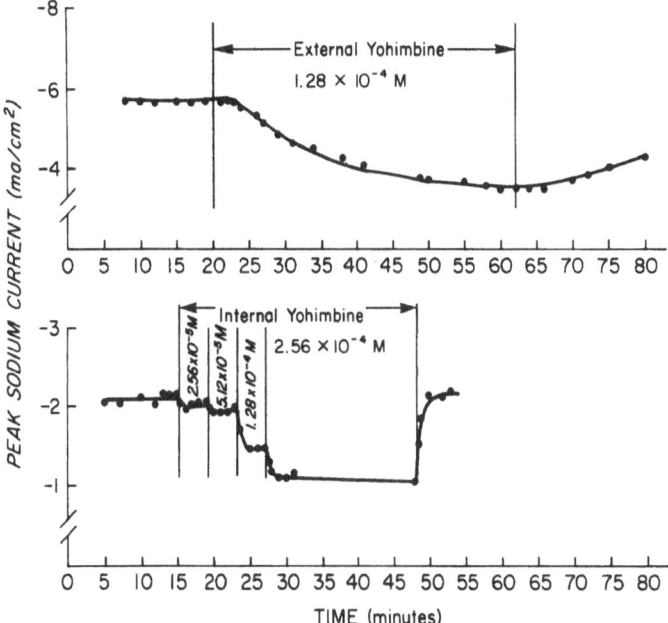

Figure 8. Time course of development of tonic effect of yohimbine in squid giant axons when applied externally (top of figure) and by internal perfusion (bottom of figure). The peak sodium current was determined with appropriate leakage subtraction. Reproduced with permission of the *Biophysical Journal*. (Lipicky *et al.*, 1978.)

positions. At pH 7, the molecule can be expected to carry a positive charge on the N at position 4. The proportion of charged molecules has not been determined nor is the electrophoretic mobility known; however, it has been totally synthesized (Stork and Nath Guthikonda, 1973) and its configuration is well defined (Ambady and Gopinath, 1973). The yohimbine we have used is commercially available and is a racemic mixture, with all components in the normal configuration.

At the time of our original studies there was no published study of the mechanism of action of yohimbine, although its use dependence, as derived from action potential analysis, was well defined (Graham, 1935; Doty and Gerard, 1950; Shanes, 1951; Holman and Shaw, 1955; Shaw *et al.*, 1955; Simon, 1955). In the following, we will present descriptive data we have obtained for yohimbine. The detail is presented in order to define clearly the observed phenomena.

Yohimbine has two phenomenologically different effects. The first effect (Figure 8) is a reduction in sodium currents when a standard test pulse is administered with a frequency no greater than 0.1 per second. However, if during the time course of development of the tonic effect (for externally or internally applied yohimbine), the axon is depolarized either by stimulating (under current-clamp conditions) or by depolarizing pulses (under voltage-clamp conditions) at frequencies greater than 0.1 per second, a further reduction of sodium current occurs (Figure 9). Thus yohimbine exerts both tonic and use-dependent effects.

A number of observations indicate that the site of action of yohimbine is more on the inner surface of the membrane than on the outer surface of the membrane. The most direct observation is shown in Figure 10. In this experiment, the reduction in sodium current produced by external application of 2.56×10^{-4} M yohimbine while the axon was being perfused with standard KF internal solution was considerably less than the reduction expected (i.e., the mean reduction from eight different axons) for this concentration of yohimbine. When the internal perfusion was stopped the sodium current fell toward the expected value, and when

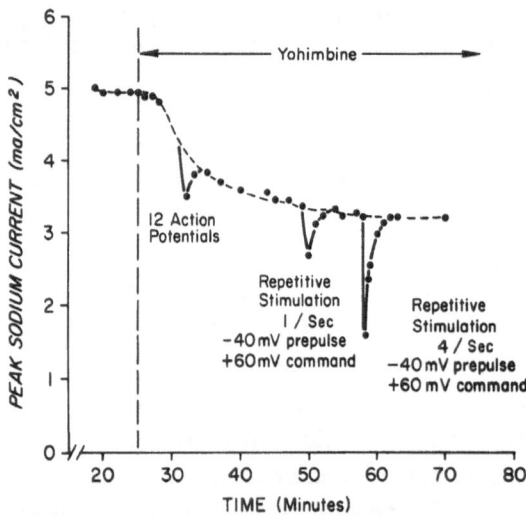

Figure 9. Time course of development of tonic effect of 1.3×10^{-4} M yohimbine when applied externally, and the additional decrease in sodium conductance when stimulated at a frequency of 1 per second or pulsed repetitively at frequencies of 1 and 4 per second. There was a bigger effect when the 4 per second frequency was used. The axon was periodically unclamped to obtain action potentials. Reproduced with the permission of the *Biophysical Journal*. (Lipicky *et al.*, 1978.)

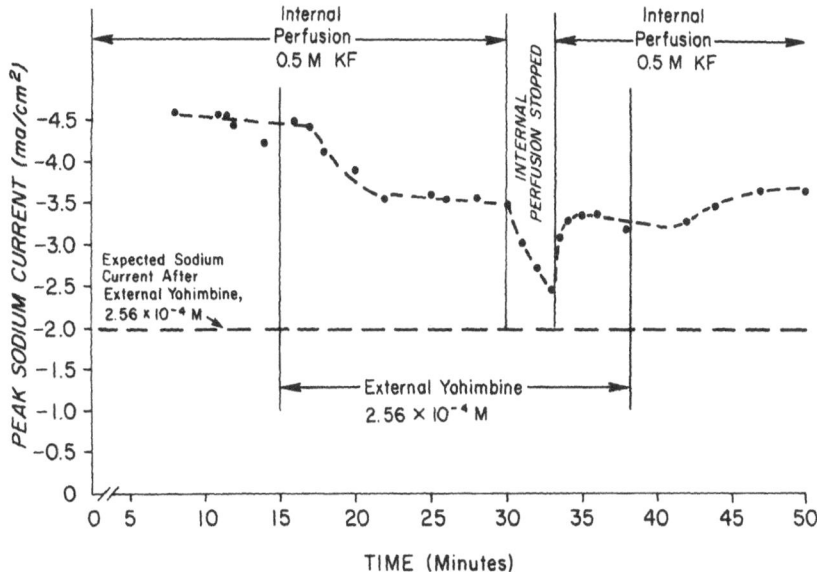

Figure 10. Time course of development of yohimbine tonic effect in squid axon when applied externally during yohimbine free internal perfusion with KF. The axon was kept clamped during the entire experiment.

the perfusion was restarted the currents increased to the previous steady-state level. This result is consistent with the action of yohimbine being dependent upon yohimbine diffusing through the membrane and acting from the inside surface. The internal perfusion in this experiment tended to dilute the internal yohimbine, thereby decreasing the effect of external application. Yohimbine, then, is like other agents (in particular local anesthetics) which also have been shown to act from the inside surface of the membrane (Narahashi *et al.*, 1969; 1970; Frazier, *et al.*, 1970; Narahashi and Frazier, 1971).

The use-dependent effect of yohimbine on the action potential is shown in Figure 11. It is of interest to note that after the fifth stimulus, even though the 100-μsec stimulus was continued at 1 per second, all that occurred was a local response (i.e, the potential change of the local response was apparently sufficient to keep use dependence at some high level). Up to the limits of solubility of yohimbine in sea water, we never have failed to see an action potential in response to the first stimulus (i.e., the tonic effect is insufficient to prevent an action potential in response to the first stimulus).

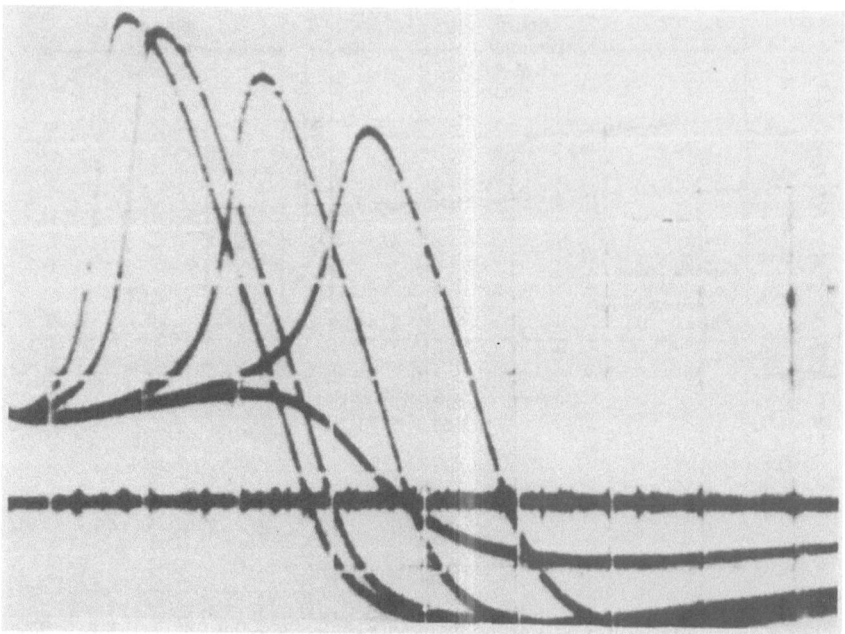

Figure 11. Effect of 2.56×10^{-4} *M* yohimbine (external) on action potentials in squid axon. The stimulus (100 μsec duration) was at 1 per second; each action potential elicited is shown; by the fifth stimulus only a local response occurred. Each horizontal unit in the grid background is 1 msec and each vertical unit is 20 mV.

The use-dependent effect of yohimbine on sodium currents is shown in Figure 12, which is a photograph of the oscilloscope trances of membrane current seen during the first, fourth, sixth, and twelfth pulses of a 1 per second train. The particular set of applied pulses was preceded by 40-msec, 40-mV hyperpolarizing (with respect to the holding potential) prepulses.

As was the case in the simulation above, the decrease of sodium currents during use-dependence with yohimbine was exponential and could be described in terms of a magnitude of reduction (at steady state) and a time constant to reach that steady state. The magnitude was increased and the time constant decreased by (a) increasing pulsing frequency, (b) increasing the concentration of yohimbine, and (c) increasing the magnitude of the depolarizing step. It is worth special note that both the magnitude and time constant of use dependence were decreased by adding hyperpolarizing pulses, preceding each command pulse.

At the time of our original investigation (Lipicky *et al.*, 1978) we expected, because of the existence of use-dependent effects, to see some obvious change in standard step-clamp measurements that would indicate that some standard Hodgkin–Huxley parameter has been altered by yohimbine. As shown in Figures 13 through 16, this was not the case. All of the data for Figures 13 through 16 were derived from step-clamp experiments where pulses were delivered at frequencies at less than 0.1 per second (i.e., representing the tonic effect).

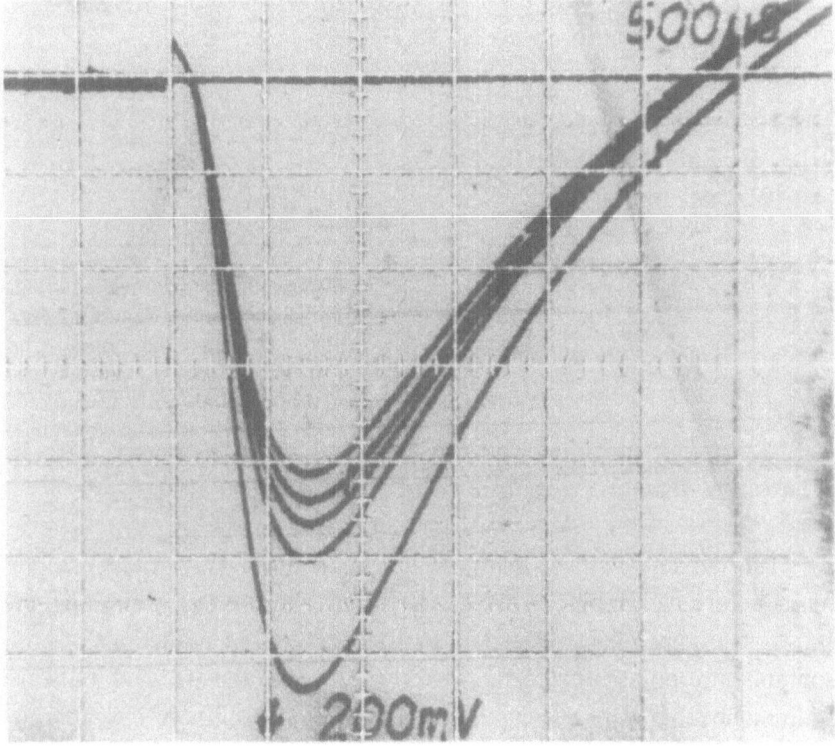

Figure 12. Use-dependent effect of 1.3×10^{-4} *M* yohimbine (external). A 20-msec, 40-mV hyperpolarizing prepulse preceded a 5-msec, 60-mV depolarizing command pulse (absolute potential, 0 mV during the command pulse), and pulses were delivered at a frequency of 1 per second. Membrane currents corresponding to pulses 1, 4, 6, 8, and 12 are shown. Each horizontal unit is 0.5 msec (shown as 500 μsec) and each vertical unit is 200 mV, which is equivalent to 0.5 mA/cm^2. Reproduced with the permission of the *Biophysical Journal*. (Lipicky *et al.*, 1978.)

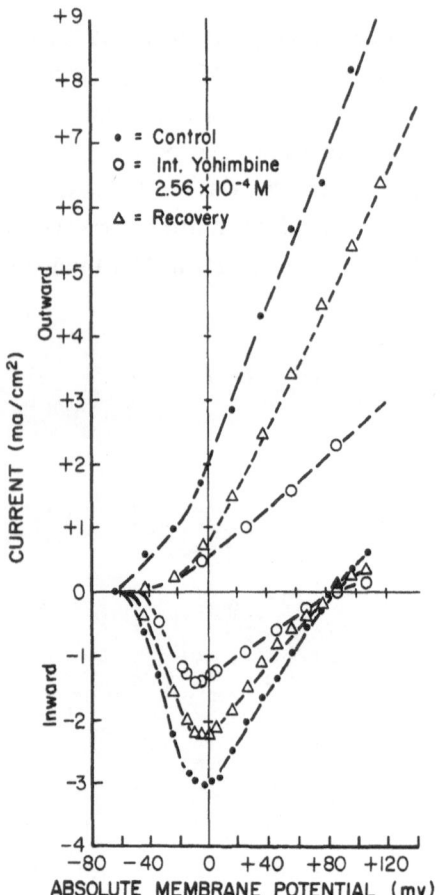

Figure 13. Current–voltage relationships during control, 2.56×10^{-4} M yohimbine (internal, in KF) and recovery. Currents all had appropriate leakage subtraction. Pulsing when in yohimbine occurred at a frequency no greater than 0.008 per second.

The current–voltage relationships for sodium and potassium currents before yohimbine, during yohimbine, and after yohimbine are shown in Figure 13. Dose related, partially reversible, reductions of both sodium and potassium currents were observed. Inward and outward sodium currents were decreased and there were no shifts in the potential at which the peak sodium current occurred. This was true for internal or external application of yohimbine and at all concentrations studied (the range of observed sodium current reduction was 5% to 60%).

Not only was there no apparent alteration of the voltage dependence of the sodium current, but also there was no qualitative alteration of the kinetics of current, except for the apparent rate of inactivation (mentioned in the simulation and discussed further below). This can be seen by inspection of the current–time records shown in Figure 12. The lack

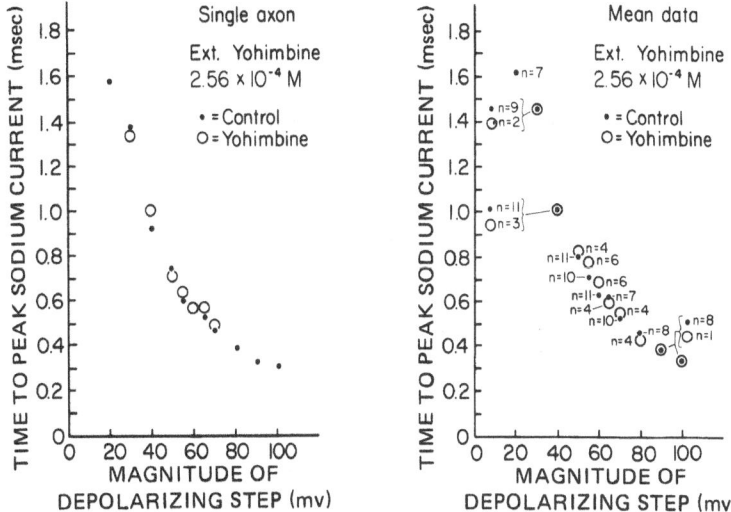

Figure 14. Relationship of the measured time-to-peak of sodium currents for each command potential. Results from a single axon are shown in the left of the figure and average values (n=number of axons represented in that point) in the right of figure.

of qualitative change in the shape of the current–time records (for the tonic effect) is demonstrated more quantitatively (and for various amplitudes of test pulse) in Figure 14 where the time-to-peak of the sodium current is plotted as a function of the magnitude of the depolarizing step.

Figure 15 shows the voltage dependence of steady-state sodium inactivation where a standard 40-msec prepulse of various voltages was

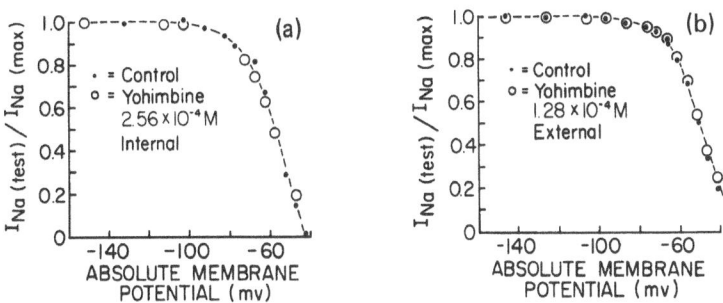

Figure 15. Voltage dependence of sodium inactivation. A standard test pulse (4 msec, 60 mV) was preceded by a 40 msec, varying magnitude prepulse; during control and 2.56×10^{-4} M yohimbine, internal (a) and 1.28×10^{-4} M yohimbine, external (b). Pulsing, when in yohimbine, occurred at a frequency no greater than 0.008 per second.

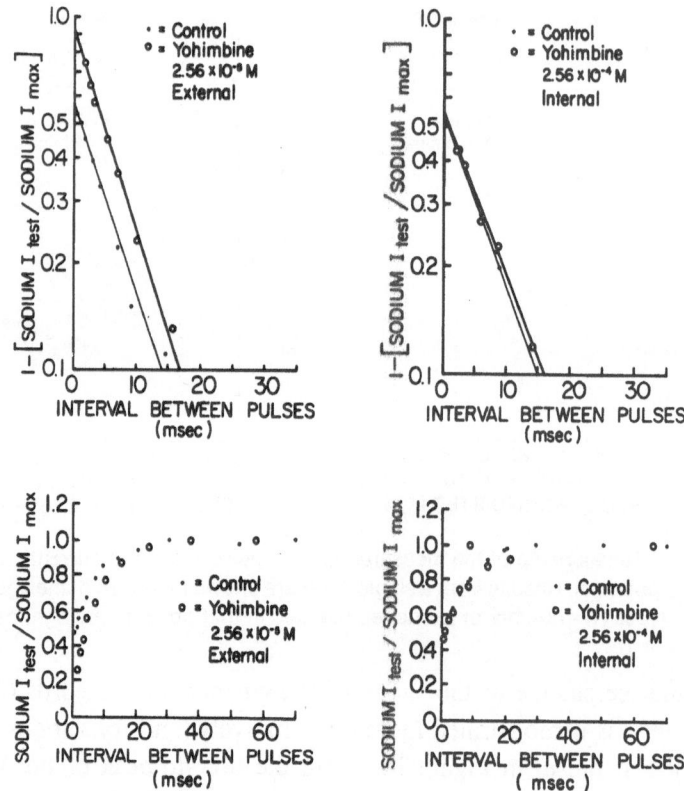

Figure 16. Time constant of inactivation during control points and 2.56×10^{-5} M yohimbine (circles) external (left panels) and 2.56×10^{-4} M yohimbine internal (right panels). Two pulses (4 msec duration, 60 mV) were delivered with varying intervals between the pulses. The bottom panels show the normalized currents. The normalized currents are equal to the maximum normalized current of one minus exp ($-$time interval/time constant). The top panels were used to obtain the time constants, which were calculated by measuring the slopes of the straight lines produced from these data.

followed by a standard depolarizing test pulse. As can be seen, there were no alterations of the voltage dependence of steady-state sodium inactivation following either internal or external application of yohimbine.

The time constant of sodium inactivation was determined in several experiments, at the resting (or holding) potential and also using a prepulse of -40 mV (with respect to the holding potential). There were no detectable differences in the time constant of sodium inactivation. The results of one experiment are shown in Figure 16.

Indeed, ignoring the observed use-dependent phenomena, yohimbine meets most standard criteria that one would apply (using step voltage-clamp methods) in order to conclude that yohimbine simply blocks the sodium channels (blockage usually meaning that the channel is straightforwardly occluded).

On first thought, provided that more drug is bound during a depolarizing pulse than at rest, one might be able to account qualitatively for the yohimbine use-dependent phenomenon, without need for more complicated schemes. That is, when yohimbine binds to the channel, the channel becomes nonconducting and yohimbine is more likely to bind if the channel is in an open conformation. However, the results of two different experimental designs have led us to conclude that this particular mechanism does not operate.

The first observations relate to the use-dependent effect as a function of command pulse duration. That is, is the use-dependent effect simply a function of potential or is it dependent upon channels being in the open state? If, for yohimbine, the use dependence was simply dependent upon potential one would expect that (a) pulsing at 1 per second with a 10-msec pulse should produce the same use-dependent reduction as pulsing at 0.1 per second with a 100-msec pulse; and (b) for any command step magnitude the use dependence should be more or less linearly related to command step duration.

Experimentally, neither prediction held. There was absolutely no equality between the use dependence produced by 0.1 per second, 100-msec pulses and 1 per second 10-msec pulses. Moreover, the relationship between command pulse duration and block (see Figure 23 in Lipicky *et al.*, 1978) showed saturation from about 2 msec through 50 msec (only 5 msec durations are shown in the Figure). Particularly from the later observations we conclude that the voltage-dependent block is not principally a function of potential. Rather, it is a reflection of potential having altered the conformation of the channel.

The other observation relates to the kinetics of the sodium current. For this experiment, we perfused the axon with a standard CsF internal solution (in order to avoid potassium current contributions to the sodium current time course). When thus perfused, the axon still produced characteristic tonic and use-dependent effects, regardless of whether yohimbine was applied internally or externally. Figure 17 compares the sodium current with internal CsF and external yohimbine after a series of

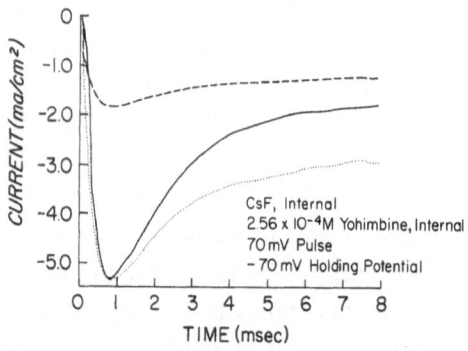

Figure 17. Two experimental sodium currents, digitized from film records (with appropriate leakage subtraction) are shown. A control current (solid line) while the squid axon was perfused with CsF and a current after addition of 2.56×10^{-4} M yohimbine to the external medium and pulsing at 1 per second until steady-state use dependence was achieved (dashed line). The magnitude of the test trace was amplified (normalized to make peak magnitudes coincide, so as to compare time sources independent of amplitude) and is the dotted line.

depolarizing pulses (maximum use-dependent effect) with the sodium current before yohimbine was added. In order to make a visual comparison, the sodium current (after steady-state use dependence in yohimbine) was normalized to the same peak magnitude as the sodium current before yohimbine (i.e., the control sodium current). The normalized current is the dotted line in Figure 17. As can be seen, inactivation for the dotted curve is smaller and occurs more slowly than in the control curve.

It is possible that the fraction of channels that do not inactivate normally (Chandler and Meves, 1970) in the presence of CsF may not undergo as much use-dependent inhibition as channels which do inactivate normally. If that is the case, it would be more appropriate to use the steady-state sodium current as the base line for normalizing. When this is done, inactivation of the normalized current appears even slower than shown in Figure 17. This same phenomenon, i.e., an apparent slowing of inactivation, occurs whether the comparison is between control and first pulse, or control and third pulse, etc.; it also occurs for nonperfused axons or axons perfused with KF or potassium glutamate. Further, dose–response relationships for the CsF steady-state sodium current are about the same as for peak sodium currents, both for tonic and use-dependent effects.

It is reasonably well established that drug binding (at least for use dependence) occurs when the channels are in the open state for yohimbine and other drugs. It is also clear that between the first and second pulse (or the third and fourth, etc., to steady state) the overall sodium conductance is decreased (i.e., total conducting channels are decreased). So the number of channels decrease during or between pulses. It seems most reasonable to conclude that the channel alteration should occur during the pulse and some form of time-dependent block should be observable. This is not the case and we therefore interpret these results as incompatible with the hypothesis that when a channel binds to drug, the channel becomes nonconducting.

That yohimbine does something other than simply occlude the sodium channel is suggested by two other observations. In neuroblastoma cells (Huang *et al.*, 1977), yohimbine was shown to be a competitive inhibitor of batrachotoxin. Batrachotoxin is known to affect gating so that sodium channels stay open, and, in this preparation, tetrodotoxin is a noncompetitive inhibitor of batrachotoxin. Thus yohimbine's competitive inhibition of batrachotoxin is consistent with yohimbine having some action other than simple occlusion. Then, in the squid axon, the voltage dependence of sodium inactivation appears to be altered during repetitive pulsing (see Lipicky *et al.*, 1978, and Ehrenstein *et al.*, 1979, for more detailed discussion). This effect is shown in Figure 18.

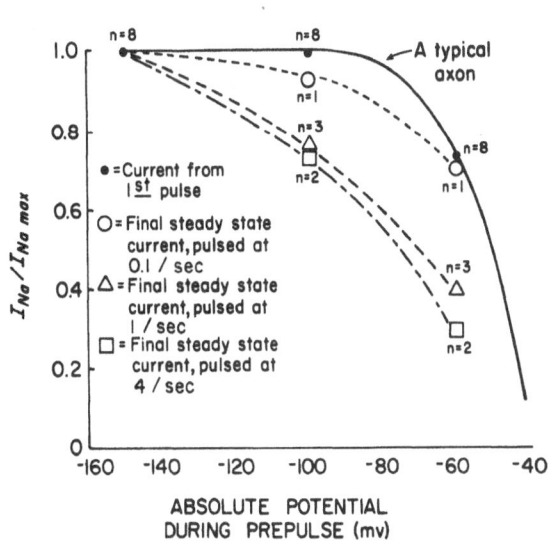

Figure 18. Steady-state inactivation curve during repetitive pulsing at 0.1 (circles), 1 (triangles), and 4 (squares) per second. The squid axon was in 1.3×10^{-4} *M* external yohimbine, prepulses were 40 msec long, and the absolute potential during the test pulse (4 msec in duration) was 0 mV.

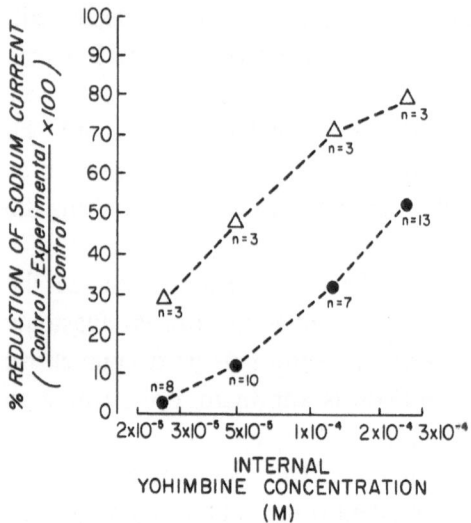

Figure 19. Dose – response curves for internal yohimbine. (O) Tonic effect. (△) Use-dependent effect. Reproduced, with minor modification, with the permission of the *Biophysical Journal*. (Lipicky *et al.*, 1978.)

As shown in Figure 19, the dose–response relationship of yohimbine concentration to sodium current inhibition (tonic or use-dependent effect) is incomplete. Because of the insolubility of yohimbine, the concentration cannot be increased sufficiently to ensure 100% occupancy of receptors (i.e., 100% drug-bound channels).

QX-314

For our purposes, assuring 100% occupancy of receptors (i.e., 100% of channels having a drug effect) is an important consideration. A simple occlusive model would predict that, at 100% occupancy, there would be no measurable current attributable to sodium channels. According to our model, at 100% occupancy, there should be a measurable current; namely, the current associated with the drug-bound-open sodium channels.

We chose to study QX-314 (structure shown in Figure 20), because of its known tonic and use-dependent effects (Frazier *et al.*, 1970; Narahashi and Frazier, 1971; Strichartz, 1973; Hille, 1977; Almers and Cahalan, 1977; Yeh, 1978; Cahalan and Almers, 1979), and because its effects on sodium conductance were known to occur at concentrations well below its solubility limits. Thus 100% occupancy of receptors could conceivably be achieved.

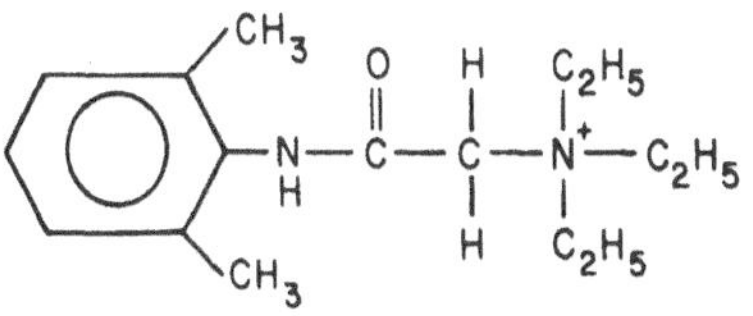

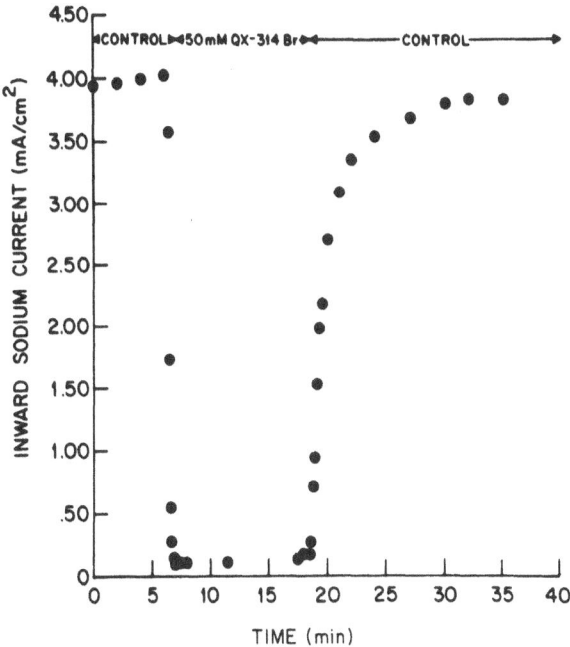

Figure 20. Structural formula for QX-314.

Like yohimbine, QX-314, when applied internally (Figure 21), produces a rapid development of a tonic effect which is completely reversible. Also, as with yohimbine, the current–voltage relationships for the
tonic reduction of sodium currents, after QX-314 (Figure 22), show no
significant alteration except for a decrease in maximum current (for both
inwardly and outwardly directed currents). The measurements of sodium
current for these purposes were made within the first two milliseconds
after the beginning of a step-depolarizing pulse. That is, the measurements were made of the early transient current, and as can be seen from

Figure 21. Time course of development of tonic effect of QX-314 when applied by
internal perfusion in a K-glutamate internal solution.

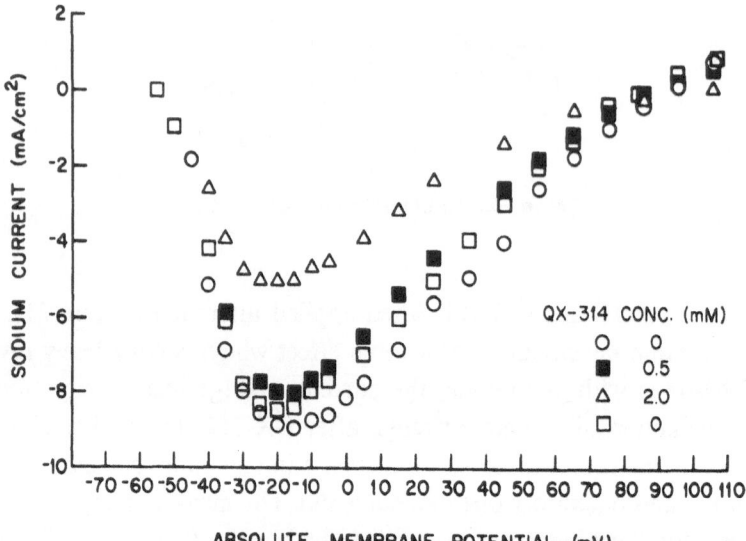

Figure 22. Current–voltage relationships for sodium currents (with appropriate leakage subtraction) during control (the first 0 concentration), 0.5 and 2.0 mM QX-314, and after recovery (the last 0 concentration). QX-314 was delivered by internal perfusion in a K-glutamate internal solution. Pulsing, when in QX-314, occurred with a frequency no greater than 0.008 per second.

Figures 23 and 24, the effect approaches 100% inhibition. This is in contrast to measurements of the same currents 10 msec or longer after the beginning of the depolarizing pulse (see below).

Figures 23 and 24 depict the tonic inhibition of the sodium currents as a function of QX-314 concentration. In Figure 23 the drug was delivered in a potassium glutamate internal solution, and in Figure 24 the drug was delivered in a standard KF internal solution of higher ionic strength. The lines through the points in both figures are least-squares fits to a standard drug receptor model (i.e., drug+receptor⇌drug receptor combination). Table II shows constants derived from these fits.

In both cases, the stochiometry indicates binding of one drug molecule to each receptor site. In other respects, the two dose–response relationships appear to be mildly different. Statistically, the intrinsic activity difference (by 2 tailed difference test with 49 degrees of freedom) gives a level of significance of 5.9%, and the intrinsic activity difference gives a level of significance of 11.9%. Although these differences are not statistically significant (at the 5% level), perhaps the differences are real.

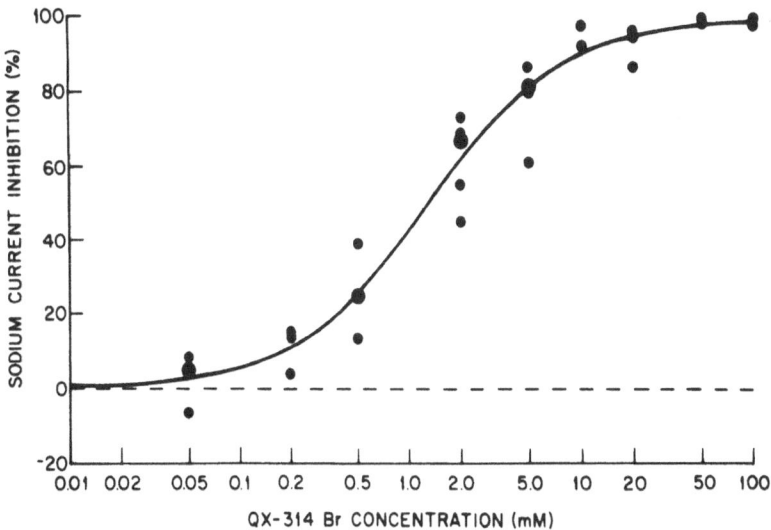

Figure 23. Dose–response relationship for sodium current inhibition when QX-314 was carried in K-glutamate internal solution.

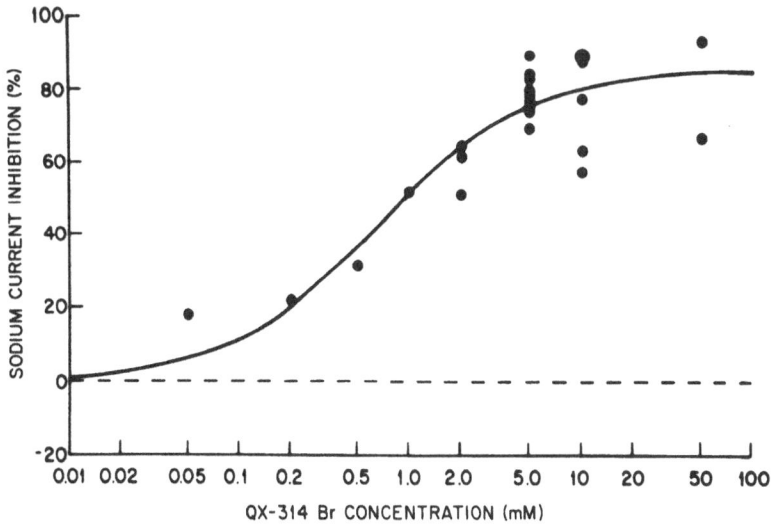

Figure 24. Dose–response relationship for sodium current inhibition when QX-314 was carried in a KF internal solution.

**Table II. Least-Squares Fit Parameters for QX-314
Dose–Response Relationships[a]**

	KF	K-Glut
Stochiometry	0.99 ± 0.28	1.08 ± 0.12
Binding constant	0.79 ± 0.10 mM	1.55 ± 0.48 mM
Intrinsic activity	$84.8 \pm 6.3\%$	$98.5 \pm 3.5\%$

[a]The intrinsic activity is equivalent to the maximum percent inhibition (Ariëns, 1954; Wand, 1968).

It would take many more observations to determine if dose–response relationships are affected by the vehicle delivering the drug. In any event, the concentration of QX-314 necessary to produce a 50% tonic sodium conductance inhibition (ED_{50}) is less than 2 mM.

The current–time relationship for any single potassium current was like those described for TEA (Armstrong, 1969; Armstrong, 1971) and quinidine (Yeh and Narahashi, 1976), in that after a peak was reached, there appeared to be a time-dependent block. The effect of QX-314 on the maximum observed potassium currents (i.e., before the droop) is shown in the potassium current–voltage relationship of Figure 25. For these maximum potassium currents, 2 mM QX-314 produced far more than a 50% inhibition. This effect of QX-314 has not been previously emphasized nor well studied. For lidocaine, procaine, and benzocaine (Taylor, 1959; Arhem and Frankenhauser, 1974), the ED_{50} for potassium currents is 10 to 100 times greater than the ED_{50} for sodium currents (i.e, sodium conductance is inhibited at much lower concentrations than is the potassium conductance). However, for QX-314, the potassium conductance is completely inhibited before the sodium conductance is completely inhibited.

For each current–time record, three measurements of potassium current were made: the maximum current, the current at 17 msec after the beginning of the step, and the current at 48 msec after the beginning of the step. Values of each of these three currents are plotted as a function of potential in Figures 25, 26 and 27. As can be seen, 2 mM QX-314 completely inhibited potassium currents, when the current was measured 48 msec after the beginning of the pulse.

Three conditions of an unambiguous experiment involving 100% occupancy of receptors are met by QX-314. First, it is readily soluble. Second, its potency for potassium conductance inhibition is greater than that for its sodium conductance inhibition; therefore, potassium current

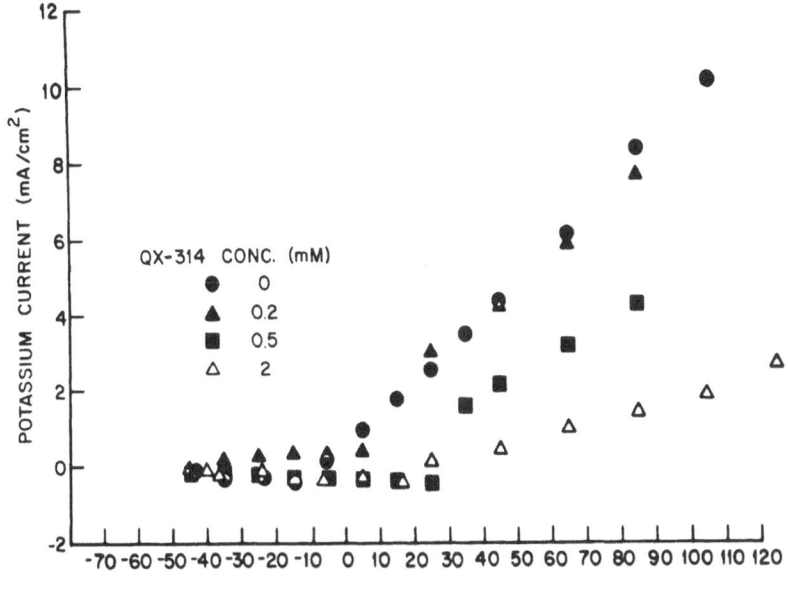

Figure 25. Current–voltage relationship for peak, outward-directed currents in control K-glutamate and in the presence of 0.2, 0.5, and 2 m*M* QX-314. Pulsing, when in QX-314, occurred with a frequency no greater than 0.008 per second.

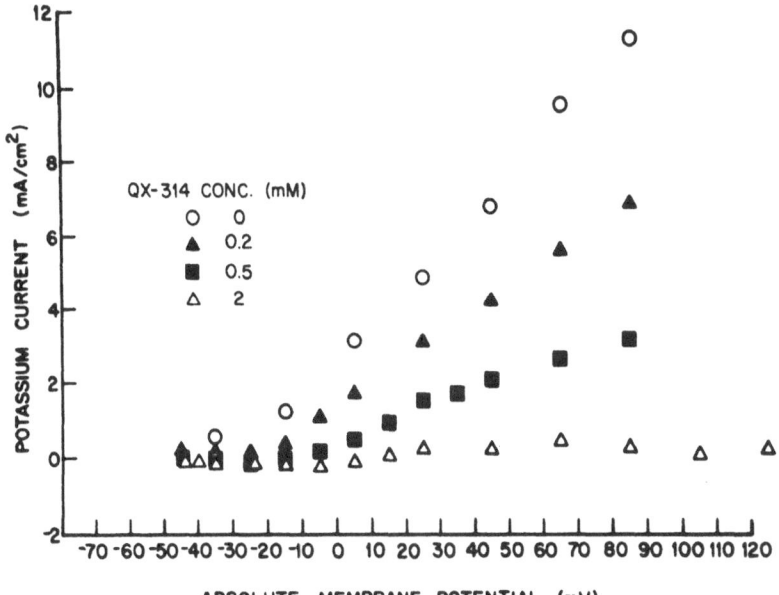

Figure 26. Current–voltage relationship for currents, measured 17 msec after the beginning of the pulse, in control K-glutamate and in the presence of 0.2, 0.5, and 2 m*M* QX-314. Same currents that are represented in Figure 25.

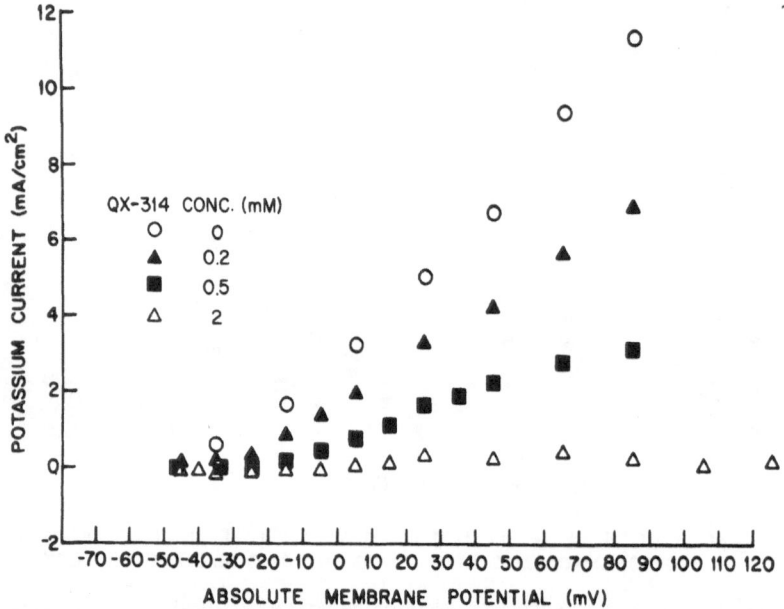

Figure 27. Current–voltage relationship for currents, measured 48 msec after the beginning of the pulse, in control K-glutamate and in the presence of 0.2, 0.5, and 2 mM QX-314. Same currents that are represented in Figure 25.

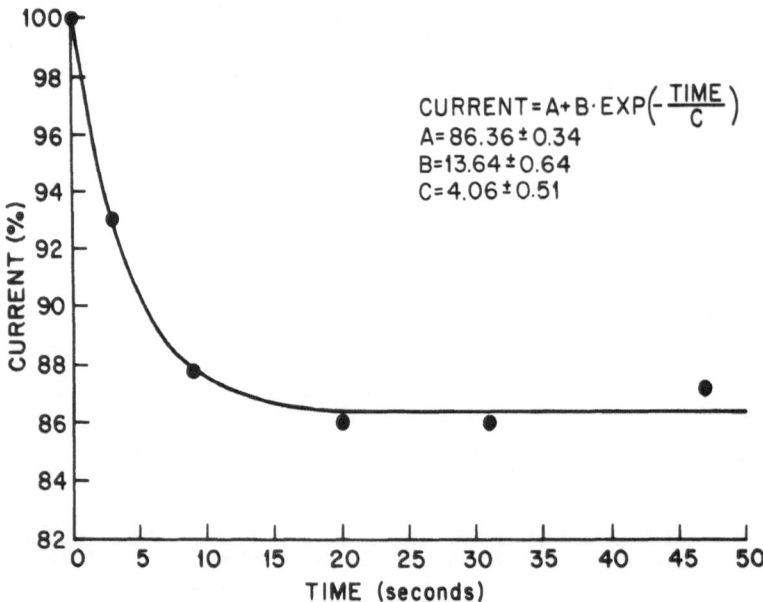

Figure 28. Use-dependent effect of QX-314. The concentration of QX-314 was 0.5 mM, internal in KF, the step pulse was 40 msec in duration, absolute potential during the pulse was −12 mV, and the pulse was delivered at 1 per second.

contributions to measured currents, can reasonably be excluded from consideration, provided that high concentrations of QX-314 (i.e., greater than 5 to 10 mM) and measurements made at long times (i.e, longer than 5 msec after the beginning of a step-depolarizing pulse) are used. Third, at less than saturating concentrations, typical tonic inhibition of sodium currents and typical use-dependent effects on sodium currents (Figure 28) were seen.

However, when the concentration of QX-314 was 10 mM or greater, inwardly directed currents were observed that were qualitatively different from the inwardly directed currents observed at lower concentrations or in the absence of QX-314. Current–time records for single pulses after 10 mM QX-314 are shown in Figures 29 through 32. These currents are the measured currents without leakage correction. As can be seen, after 10 mM QX-314 the early, transient (presumably sodium) current deviates markedly from its expected shape. As mentioned above, with 10 mM QX-314, there are no normal potassium currents distorting the early transient current recording (i.e., one should be observing only sodium channel behavior).

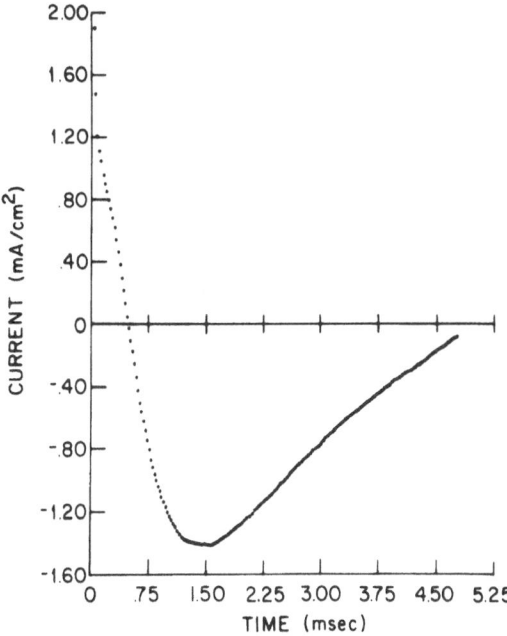

Figure 29.   A current record reproduced by digitizing a film record. Pulse magnitude was 60 mV, internal solution was KF (leakage has not been subtracted).

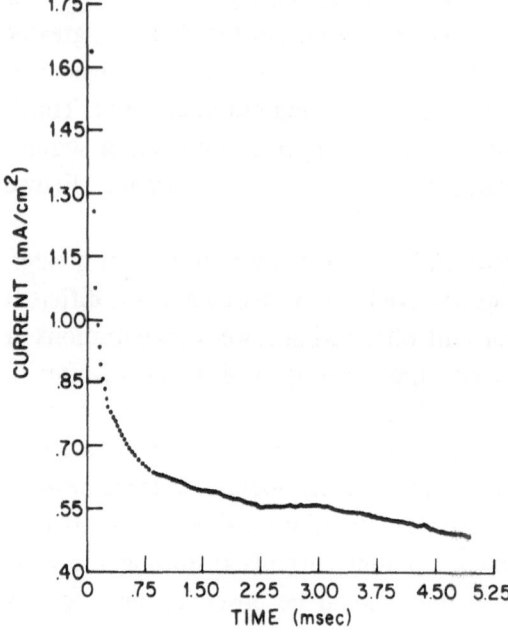

Figure 30. A current record reproduced by digitizing a film record. Pulse magnitude was 60 mV, internal solution was KF (leakage has not been subtracted). The record was obtained 0.5 min after beginning 10 mM QX-314; and about 2 min after the record shown in Figure 29.

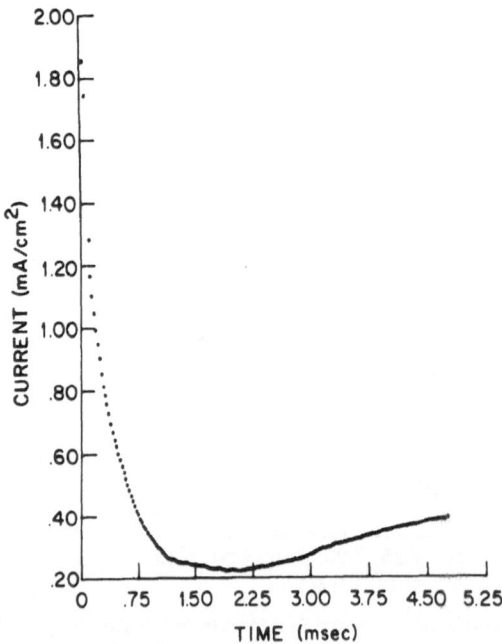

Figure 31. A current record reproduced by digitizing a film record. Pulse magnitude was 60 mV (leakage has not been subtracted). The record was obtained 0.4 min after removal of QX-314 from the KF internal solution.

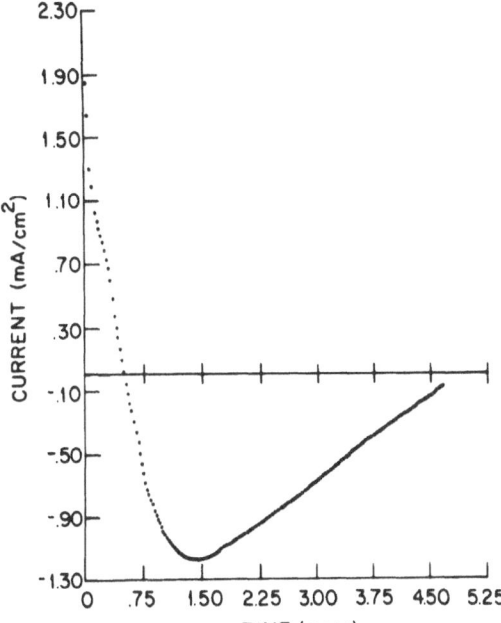

Figure 32. A current record reproduced by digitizing a film record. Pulse magnitude was 60 mV (leakage has not been subtracted). The record was obtained 3.6 min after removal of QX-314 from the KF internal solution; and 3.2 min after the record shown in Figure 31.

At a concentration of 50 mM QX-314 the voltage–current relationship shown in Figure 33 was obtained. Before QX-314, a normal-appearing control sodium current was observed. After 50 mM QX-314, the current–voltage relationship observed depended upon how long after the beginning of the pulse the measurement was made. If the measurement was made about 4 msec after the beginning of the pulse, inwardly directed currents were negligible. However, if the measurements were made about 820 msec after the beginning of a long pulse, there were appreciable inwardly directed currents. The same leakage correction was made for the 4-msec measurement as for the 820-msec measurement. The two measurements were made from the same trace for each pulse.

It appears from these observations, that at high concentrations (i.e., from 5 to 50 mM QX-314) there are no longer any "normal" sodium or potassium currents. In their place are inwardly directed currents that have long time courses (reaching peaks in hundreds of milliseconds) and their magnitudes increase with increasingly large depolarizing potentials.

This phenomenon was observed in 13 out of 15 exposures to QX-314, in potassium glutamate (14 axons; 3 exposures at 2 mM, 4 at 5

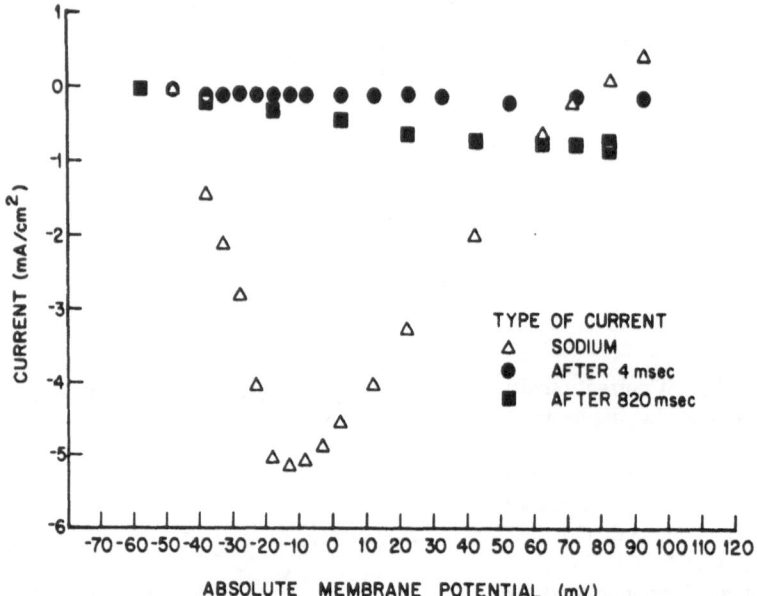

Figure 33. Current–voltage relationships for normal, control sodium currents and for currents measured 4 and 820 msec after the beginning of the pulse while the axon was exposed to 50 m*M* QX-314 in a K-glutamate internal solution.

m*M*, 2 at 10 m*M*, 3 at 20 m*M* and 2 at 50 m*M*). The two "failures" were in 2 m*M* QX-314. In 2 axons, when external sodium was replaced by Tris, this unusual inwardly directed current disappeared. Consequently, we assume the current was being carried by sodium ions. In eight experiments, TTX at concentrations of 1×10^{-7} *M* reduced the inward current by about 10% in four axons, and had no effect on this current in the other four axons.

These currents were small and their absolute magnitudes were highly dependent upon the leakage correction. Records were made by photographing an oscilloscope screen so leakage corrections were rather imprecise. Therefore, there is some uncertainty in the magnitude, time course, and voltage dependence of these currents. However, in 2 axons exposed to Tris sea water we did not detect any inwardly directed currents, using the same leakage correction method.

Within the these limits of uncertainty, because of the high solubility of QX-314 and because it is more potent as a potassium current inhibitor than as a sodium current inhibitor, it was possible to obtain evidence for a drug-bound channel whose kinetics are different from kinetics observed for a normal sodium channel. We interpret these results to be consistent with our model and to be inconsistent with a simple occlusion model.

DISCUSSION

Since the initial description of use-dependent phenomena (Strichartz, 1973) a variety of drugs have been shown to have use-dependent properties. All of the drugs which exhibit use dependence also have tonic effects. Among these agents are local anesthetics, such as lidocaine, QX-314, and other derivatives which share a common structure. However, other agents such as yohimbine, quinidine, and strychnine, although exhibiting use dependence, have no structural similarity to local anesthetics or to one another, and have widely different biological actions. Tetrodotoxin, yet another chemical structure, inhibits sodium conductance, but exhibits no use dependence; thus, inhibiting sodium conductance does not in itself necessarily imply use dependence. Since diverse pharmacological classes of drugs produce use dependence, there is no compelling reason to assume a common mechanism of action nor to assume that a mechanism shown for strychnine is also an explanation for the action of QX-314, lidocaine, or yohimbine.

General models accounting for use dependence have been formulated. Early ideas were heavily oriented toward a basic occlusion mechanism (see Hille, 1977), but certain measures of channel kinetic alteration were necessary in order to make these models fit with the data. More recently, sodium inactivation has been strongly implicated in use-dependent phenomena. Support for this idea comes from observations showing that after pronase pretreatment, use dependence is absent for QX-314 (Almers and Cahalan, 1977; Yeh, 1978; Cahalan, 1978), strychnine (Cahalan and Shapiro, 1976), N-methylstrychnine, QX-222 (Cahalan, 1978), and 9-α-amino-acridine (Yeh and Narahashi, 1976; Cahalan, 1978). The same phenomenon is seen after other methods for altering sodium inactivation are used (Courtney, 1975; Hille et al., 1975; Khodorov, et al., 1976; Hille, 1977).

At present there appears to be no experimental result which necessitates invoking a simple occlusive mechanism for any drug that exhibits use dependence. In fact, the experimental results necessitate mechanisms other than simple occlusion. For example, recently Cahalan and Almers (1979, 1979b), in order to account for experimental results, found it necessary to postulate two receptors and two mechanisms of action (one for the tonic effect and one for the use-dependent effect of QX-314). Moreover, they interpret their results to indicate that a drug-bound channel is equivalent to an inactivated channel. The idea of two receptors for local anesthetics has independent support from a study of sodium transport in neuroblastoma tissue culture (Huang and Ehrenstein, 1981). The models and considerations are thus becoming more complex.

The evidence from the pronase experiments cited above strongly supports a role of inactivation in the use-dependent phenomena, since pronase clearly removes sodium inactivation (Armstrong *et al.*, 1973; Rojas and Armstrong, 1971; Rojas and Rudy, 1976). In performing pronase experiments, it was necessary to also block potassium currents (by perfusing with either CsF or TEA). Cahalan and Almers have shown an interaction between tetrodotoxin and QX-314 on gating currents. It is possible that interaction between TEA, CsF, and QX-314 (or other agents) exists after pronase treatment. Our experiments with yohimbine in CsF and those of Yeh (1978) for QX-314 in CsF show use dependence. Yet in CsF there is incomplete inactivation of some sodium channels.

It is clear that inhibition of sodium currents can conceivably be caused by occlusion of the sodium channel or by changing the kinetic characteristics of the sodium channel. To account for the use-dependent phenomena, other investigators have had to include in their models both of these alterations and/or multiple binding sites for drug action. Our model has the distinct advantage of assuming only a change in the kinetic characteristics and only one binding site. We believe it is not necessary to assume occlusion for use-dependent effects caused by all drugs.

Intuitively, it is reasonable to suppose that the sodium channel in its open state offers binding sites that are otherwise not readily available and that these sites might be either in the lumen or along the channel–lipid interface. Because even the resting membrane has a sodium channel turnover, the drug might bind only to open channels and result in a tonic inhibition. This is amply demonstrated by our simulation. Given a drug–channel combination, it is equally reasonable to suppose that the voltage dependence or time dependence of the channel behavior might be altered. We have shown by simulation that simply altering one Hodgkin–Huxley parameter leads to typical use dependence. Thus, it is not necessary to postulate a receptor or receptors that are different for the resting axon as opposed to the active axon, nor to postulate a different mechanism for the tonic effect as opposed to the use-dependent effect.

SUMMARY

We have presented some aspects of our developing model of drug action which can explain use-dependent phenomena in excitable membranes. In addition, we have presented data from the study of two drugs of completely different structure, which exhibit use-dependent inhibition of ionic channels, and we have discussed briefly how these experimental results are consistent with our model. Our model is not meant to account for all drug effects (in fact, our intuition leads us to think that it is not possible to account even for a single class of drugs by one mechanism). It does appear to account for observations we have made with yohimbine and QX-314.

ACKNOWLEDGMENTS

We thank Freda Jacobsen, Stephan Grupp, Liz Wendelmoot, Thomas Corner, Maxine Schaefer and Edward Grood for their help in this work. Gerald Ehrenstein has contributed much discussion. R. J. Lipicky was supported, in part, by N. I. H. grant No. NS-12635.

REFERENCES

Almers, W., and Cahalan, M. D. (1977). Interaction between a local anesthetic, the sodium channel gates and tetrodotoxin, *Biophys. J.* **17**, 205a (Abstr).

Ambady, G., and Gopinath, K. (1973). Crystal structure and absolute configuration of yohimbine hydrochloride, $C_{21}H_{27}ClN_2O_3$, *J. Cryst. Mol. Struct.* **3**, 37–45.

Århem, P., and Frankenhaeuser, B. (1974). Local anesthetics: Effects on permeability properties of nodal membrane in myelinated nerve fibres from Xenopus. Potential clamp experiments, *Acta Physiol. Scand.* **91**, 11–21.

Ariëns, E. J. (1954). Affinity and intrinsic activity in the theory of competitive inhibition, *Arch. Int. Pharmacodyn.* **99**, 32–49.

Armstrong, C. M. (1969). Inactivation of the potassium conductance and related phenomena caused by quaternary ammonium ion injection in squid axons, *J. Gen. Physiol.* **54**, 553–575.

Armstrong, C. M. (1971). Interaction of tetraethylammonium ion derivatives with the potassium channels of giant axons, *J. Gen. Physiol.* **58**, 413–437.

Armstrong, C. M., Bezanilla, F., and Rojas, E. (1973). Destruction of sodium conductance inactivation in squid axons perfused with pronase, *J. Gen. Physiol.* **62**, 375–391.

Cahalan, M. D. (1978). Local anesthetic block of sodium channels in normal and pronase-treated squid giant axons, *Biophys. J.* **23**, 285–311.

Cahalan, M. D., and Almers, W. (1979). Interactions between quaternary lidocaine, the sodium channel gates, and tetrodotoxin, *Biophys. J.* **27**, 39–56.

Cahalan, M. D., and Almers, W. (1979b). Block of sodium conductance and gating current in squid giant axons poisoned with quaternary strychnine, *Biophys. J.* **27**, 57–74.

Cahalan, M. D., and Shapiro, B. I. (1976). Current and frequency dependent block of sodium channels by strychnine, *Biophys. J.* **16**, 76a (Abstr).

Chandler, W. K., and Meves, H. (1970). Evidence for two types of sodium conductance in axons perfused with sodium fluoride solution, *J. Physiol. (Lond.)* **211**, 653–678.

Cole, K. S. (1965). *Membranes Ions Impulses* (University of California Press, Berkeley and Los Angeles, California).

Courtney, K. R. (1975). Mechanism of frequency-dependent inhibition of sodium currents in frog myelinated nerve by the lidocaine derivative GEA 968, *J. Pharmacol. Exp. Ther.* **195**, 225–236.

Doty, R. W., and Gerard, R. W. (1950). Nerve conduction without increased oxygen consumption: action of azide and fluoroacetate, *Am. J. Physiol.* **162**, 458–468.

Ehrenstein, G., Lipicky, R. J., and Gilbert, D. L. (1979). Effects of use dependent inhibition by yohimbine on sodium inactivation. *Biophys. J.* **25**, 135a (Abstr.)

Frazier, D. T., Narahashi, T., and Yamada, M. (1970). The site of action and active form of local anesthetics. II. Experiments with quaternary compounds, *J. Pharmacol. Exp. Ther.* **171**, 45–51.

Graham, H. T. (1935). The subnormal period of nerve response, *Am. J. Physiol.* **111**, 452–465.

Hamet, R. (1925). Sur un nouveau cas d'inversion des effets adrénaliniques, *C. R. Acad. Sci.* **180**, 2074–2077.

Hille, B. (1977). Local anesthetics: hydrophilic and hydrophobic pathways for the drug-receptor reaction, *J. Gen. Physiol.* **69**, 497–515.

Hille, B., Courtney, K., and Dum, R. (1975). In *Molecular Mechanism of Anesthesia*, B. R. Fink, Ed., *Progress in Anesthesiology*, Vol. I (Raven Press, New York), pp. 13–24.

Hodgkin, A. L., and Huxley, A. F. (1952). A quantitative description of membrane current and its application to conduction and excitation in nerve, *J. Physiol.* **117**, 500–544.

Holman, M. E., and Shaw, F. H. (1955). The effect of yohimbine and other drugs on the isolated frog skin potential, *Aust. J. Exp. Biol. Med. Sci.* **33**, 671–676.

Huang, L. M., and Ehrenstein, G. (1981). Local anesthetics QX 572 and benzocaine act at separate sites on the batrachotoxin-activated sodium channel, *J. Gen. Physiol.* **77**, 137–153.

Huang, L. M., Ehrenstein, G., and Catterall, W. A. (1978). Interaction between batrachotoxin and yohimbine, *Biophys. J.* **23**, 219–233.

Khodorov, B., Shishkova, E., Peganov, E., and Revenko, S. (1976). Inhibition of sodium currents in frog Ranvier node treated with local anesthetics. Role of slow sodium inactivation. *Biochim. Biophys. Acta* **433**, 409–435.

Lipicky, R. J., Gilbert, D. L., and Ehrenstein, G. (1978). Effects of yohimbine on squid axons, *Biophys. J.* **24**, 405–422.

Narahashi, T., and Frazier, D. T. (1971). Site of action and active form of local anesthetics. In *Neurosciences Research*, S. Ehrenpreis and O. C. Solnitzky, Eds. Vol. 4 (Academic, New York), pp. 65–99.

Narahashi, T., Frazier, D. T., and Yamada, M. (1970). The site of action and active form of local anesthetics. I. Theory and pH experiments with tertiary compounds, *J. Pharmacol. Exp. Ther.* **171**, 32–44.

Narahashi, T., Moore, J. W., and Poston, R. N. (1969). Anesthetic blocking of nerve membrane conductances by internal and external applications, *J. Neurobiol.* **1**, 3–22.

Neher, E., and Steinbach, J. H. (1978). Local anesthetics transiently block currents through single acetylcholine-receptor channels, *J. Physiol. (Lond.)* **277**, 153–176.

Nickerson, M. (1949). The pharmacology of adrenergic blockade, *Pharmacol. Rev.* **1**, 27–101.

Rojas, E., and Armstrong, C. M. (1971). Sodium conductance activation without inactivation in pronase-perfused axons, *Nat. New Biol.* **229**, 177–178.

Rojas, E., and Rudy, B. (1976). Destruction of the sodium conductance inactivation by a specific protease in perfused nerve fibres from *Loligo*, *J. Physiol. (Lond.)* **262**, 501–531.

Schwarz, W., Palade, P. T., and Hille, B. (1977). Local anesthetics: Effect of pH on use-dependent block of sodium channels in frog muscle, *Biophys. J.* **20**, 343–368.

Seyama, I., Wu, C. H., and Narahashi, T. (1980). Current-dependent block of nerve membrane sodium channels by paragracine, *Biophys. J.* **29**, 531–537.

Shanes, A. (1951). Electrical phenomena in nerve. III. Frog sciatic nerve, *J. Cell. Comp. Physiol.* **38**, 17–40.

Shapiro, B. I. (1977). Effects of strychnine on the sodium conductance of the frog node of Ranvier, *J. Gen. Physiol.* **69**, 915–926.

Shapiro, B. I., Wang, C. M., and Narahashi, T. (1974). Effects of strychnine on ionic conductances on squid axon membrane, *J. Pharmacol. Exp. Ther.* **188**, 66–76.

Shaw, F. H., Holman, M., and Mackenzie, J. G. (1955). The action of yohimbine on nerve and muscle of amphibia, *Aust. J. Exp. Biol. Med. Sci.* **33**, 497–505.

Simon, S. E. (1955). The effect of yohimbine on sodium and potassium movements in resting nerve and muscle, *Aust. J. Exp. Biol. Med. Sci.* **33**, 179–188.

Stork, G., and Nath Guthikonda, R. (1972). Stereoselective total synthesis of ($\pm$)-yohimbine, ($\pm$)-Ψ-yohimbine, and ($\pm$)-β-yohimbine, *J. Am. Chem. Soc.* **94**, 5109–5110.

Strichartz, G. (1973). The inhibition of sodium currents in myelinated nerve by quaternary derivatives of lidocaine, *J. Gen. Physiol.* **62**, 37–57.

Swenson, R. P., Oxford, G. S., and Narahashi, T. (1978). Enhancement of sodium channel inactivation by octanol and decanol, *Biophys. J.* **21**, 41a (Abstr.).

Taylor, R. E. (1959). Effects of procaine on electrical properties of squid axon membrane, *Am. J. Physiol.* **196**, 1071–1078.

Wand, D. R. (1968). Pharmacological receptors, *Pharmacol. Rev.* **20**, 49–88.

Werner, G. (1958). Zur Wirkung von "Rauwolfi serpentina," *Arznein.-Forsch.* **4**, 40–41.

Yeh, J. Z. (1978). Sodium inactivation mechanism modulates QX-314 block of sodium channels in squid axons, *Biophys. J.* **24**, 569–574.

Yeh, J. Z., and Narahashi, T. (1976). Mechanism of action of quinidine on squid axon membranes, *J. Pharmacol. Exp. Ther.* **196**, 62–70.

Yeh, J. Z., and Narahashi, T. (1977). Kinetic analysis of pancuronium interaction with sodium channels in squid axon membranes, *J. Gen. Physiol.* **69**, 293–323.

Index